R.S. Means Company, Inc.

INSURANCE REPAIR:

Opportunities, Procedures and Methods

Peter J. Crosa, AIC

R.S. MEANS COMPANY, INC.
CONSTRUCTION CONSULTANTS & PUBLISHERS
100 Construction Plaza
P.O. Box 800
Kingston, MA 02364-0800
(617) 585-7880
© 1989

This book was edited by Allan Cleveland, Mary Greene, and Julia Willard. Typesetting was supervised by Joan Marshman. The book and jacket were designed by Norman Forgit. Illustrations by Carl Linde.

Printed in the United States of America.

10   9   8   7   6   5   4   3   2   1
Library of Congress Cataloging in Publication Data
ISBN 0-87629-146-9

This book is dedicated to Melissa.

TABLE OF CONTENTS

FOREWORD

Billions of dollars are paid out annually by insurance companies to restore damaged property. Most of that money is paid to general contractors, carpenters, plumbers, electricians, painters, and carpet layers. While insurance repair work represents a tremendous source of income for these contractors, their availability to perform this work is equally important to the insurance industry.

> *A homeowner finds that a windstorm has blown shingles off of his roof. He calls his insurance company. The insurance company calls a roofer.*
>
> *Fire destroys a warehouse. The owner calls his insurance company. The insurance company calls a contractor.*
>
> *Lightning strikes a butcher shop, blowing out the freezers and jeopardizing the company's inventory and business. The butcher calls his insurance company. The insurance company calls a refrigerator mechanic.*

The fact is, accidents happen, regardless of the state of the economy. Therefore, insurance repair jobs represent a steady source of income.

This book is a step-by-step guide to show a skilled craftsman or contractor how to start up, operate, and expand a business restoring damaged buildings, furnishings, or other property for insurance companies. It explains insurance company operations and discusses how contractors can successfully sell their services to insurance adjusters.

Chapter 1 provides a background of the roles and relationships of the contractor, the insurance adjuster, and the policyholder. The image and marketing of the insurance repair contractor is the subject of Chapter 2.

Where and how does the contractor meet insurance representatives with whom to form liaisons? Establishing these all-important contacts is the key to successful insurance repair contracting. It is the adjusters who will provide the contractor with jobs. This topic is addressed in Chapter 3.

Writing a proper estimate can be as important to getting a job as making contacts with adjusters. Insurance repair contractors must write estimates according to standards set by the insurance industry; these specific procedures are explained in Chapter 4.

The "bottom line" in insurance repair contracting is *getting paid*. Who holds the insurance company's check? How much will the policyholder be expected to pay? Chapter 5 suggests methods the contractor can use to make sure he receives compensation in a timely manner for the full amount of the claim.

The last two chapters are sample estimates – one residential and one commercial. In these chapters, the proper procedures for reporting and writing quantity takeoffs and estimates are demonstrated and explained. Sample takeoff and estimate sheets, as well as other correspondence, are included to familiarize contractors with the requirements of an actual project.

The extensive Appendix contains names and addresses of insurance associations, samples of letters to use for collection and introduction, and a format contractors can use to produce their own individualized unit price guide. Finally, a glossary of insurance terms provides a useful reference for those entering the insurance repair field.

PREFACE

During 20 years as an insurance claims adjuster, I have signed insurance company checks totalling millions of dollars made payable to general contractors for the restoration of damaged property. I have handled insurance claims throughout the Eastern Seaboard, from Florida to the District of Columbia. Over the years, I've met many builders who wanted to know how they could get into the business of insurance repair. They knew their particular building trade well, and had already proven that there was a good living to be made in new construction and remodeling. They also knew that there was a way to expand their business, with an extraordinarily wealthy new customer, *the insurance company*.

In order to succeed in the business of insurance repair contracting, one must not only be proficient in a particular building trade, but must also understand the insurance repair industry and how insurance company restoration money is paid out to contractors (often in advance). Without this knowledge, contractors may be missing out on some very lucrative opportunities.

This book was written in answer to the contractors who questioned me about the opportunities in insurance repair. It condenses my many years of practice as an insurance claims adjuster into a guide for the contractor who wishes to enter the insurance repair field. The book covers issues ranging from where to find and how to contact insurance companies and their claims adjusters, to how these adjusters operate. This includes how insurance companies make decisions about spending insurance company dollars, how they choose repair contractors, and what they expect from the repair contractor.

Peter J Crosa

INTRODUCTION: HOW TO USE THIS BOOK

The chapters in this book are written to correspond with a step-by-step process. The information in each chapter provides the foundation needed to best utilize the information presented in the next chapter. The significance of this chronology and the relationship of chapters are explained below.

Chapter 1 provides an overview of the roles of insurance companies, claims adjusters, and policyholders. This overview is a place to start, to consider the venture of insurance repair. A successful hunter first learns the nature of the beast he hunts. What are its habits? What are its needs? It is only when he understands the motivation of the beast that he is able to position himself to "capture" the prize.

Although each insurance company has its own philosophy and method of doing business, it is possible to acquire an understanding of the objectives of insurance companies in general. By understanding its objectives, the contractor can better visualize its needs. (Later chapters in this book advise the contractor on how to position himself to fill those needs.)

The same applies to the objectives and needs of the insurance claims adjuster and the policyholder. The needs of the three parties are distinct and, at times, opposed. The contractor interested in insurance repair work should understand the particular requirements of each of these parties.

Chapter 2 closely examines the role of the successful insurance repair contractor. To begin with, he must be perceived by claims adjusters as a professional. This image must be maintained in all aspects of the contractor's operation, from proper licensing, bonding, and insurance, to the services he provides and the prices he charges. Most insurance adjusters have had enough exposure to contractors to know which are the successful, reliable professionals and which are the "fly-by-nights." What is the condition of the company vehicles, office, or warehouse? Do employees reflect professional behavior and appearance? These considerations should be reviewed before a contractor even approaches the insurance industry.

Where and how can contractors establish contact with insurance company claims representatives? How does one get the opportunity to restore damaged property? Chapter 3, entitled "Getting the Attention of Claims Adjusters" focuses on how to approach the insurance industry.

Chapter 4 proceeds to the next step, where the contractor actually begins fulfilling some of the needs of the insurance industry – writing repair estimates for insurance claims adjusters. Keep in mind that the average claims adjuster is neither a skilled contractor nor an engineer. The objective is to prepare an estimate that most accurately reflects job costs, and that is prepared according to the methods and standards preferred by insurance agencies.

The ultimate purpose of entering the insurance repair field is to make a profit. If there is a negative aspect to this lucrative business, it is working out the finances. Many insurance repair contractors undertake restoration work and finish the job, only to discover that the settlement check has been sent to a mortgage bank 3,000 miles away. However, there are ways to ensure that financial matters will be clear and in order before a project is begun. Insurance repair contractors must understand the complex path that insurance money follows. Chapter 5 provides detailed guidance on how to secure payment, estimates, and bonds.

Chapters 6 and 7 present two sample estimates – one residential and one commercial. These examples serve as models of the required paperwork, showing exactly how to set up and present these documents to the insurance adjuster.

The Appendix contains useful references and sample organizational formats for the contractor setting up an insurance repair business. Included is a unit price list that can be tailored for use in a contractor's particular area. There are also sample letters of introduction that contractors can use as a tool to get into the market. Sample collection letters are included for use on those occasions when payment is not secured as planned. These notices are written with the force of an attorney, but without the expense.

Finally, a glossary of insurance terms contains a basic core of insurance claims terms that the contractor will find helpful in communicating with claims adjusters. One might acquire this vocabulary after years in the insurance repair business, but having these definitions up-front can save time and avoid confusion.

How Much Money Can You Earn?

Billions of dollars are paid annually by insurance companies for the repair of damaged structures nationwide. It may not be easy to visualize one billion dollars, but consider: If you were to count one billion single dollar bills, at the rate of one per second, you would be well into your 32nd year before you counted the billionth single dollar bill.

With a billion dollars you could build $100,000 homes for 10,000 families. You could build 5-million dollar schools for 200 school districts across your state. You could build 10-million dollar hospitals for 100 communities across your state.

Insurance companies have paid over 50 billion dollars per year to restore damaged property throughout this country. Most of that money went to general contractors, carpenters, plumbers, electricians, painters, carpet layers; you name the construction skill, and an insurance company has paid for it.

The figure below is a chart compiled by the A. M. Best Co. listing the expenditures of the insurance industry on claims involving commercial building and dwelling coverages. All but a small portion of this money was applied to the reconstruction of damaged property.

The amount depends on the type and size of a contracting service. Success stories range from one-man operations to large contracting corporations.

Success Formula

Understanding the elements and workings of the insurance industry is only a part of the success formula. Building an image as a professional insurance repair contractor and winning the confidence of claims adjusters and policyholders is another major

Operating Results, Property/Casualty Insurers, 1984–1986
(000 omitted from dollar figures)

	1986	1985	1984	% Change 1985-86	% Change 1984-85
Assets	$374,088,083	$311,364,665	$264,734,722	+ 20.1%	+ 17.6%
Net Premiums Written	176,552,070	144,186,420	118,166,311	+ 22.4	+ 22.0
Premiums Earned	166,381,186	133,341,852	115,009,836	+ 24.8	+ 15.9
Losses and Adjustment Expenses Incurred	135,765,622	118,572,435	101,445,723	+ 14.5	+ 16.9
Loss and Loss Adjustment Ratio	81.60%	88.92%	88.21%	− 8.2	+ 0.8
Underwriting Expenses Incurred	44,542,343	37,585,418	33,184,338	+ 18.5	+ 13.3
Underwriting Expense Ratio	25.23%	25.95%	27.98%	− 2.8	− 7.3
Combined Ratio (before Dividends to Policyholders)	106.83%	114.87%	116.19%	− 7.0	− 1.1
Combined Ratio (after Dividends to Policyholders)	108.13%	116.52%	118.01%	− 7.2	− 1.3
Statutory Underwriting Gain (Loss)	(13,747.925)	(22,597,292)	(19,378,885)	− 39.2	+ 16.6
Dividends to Policyholders	2,164,969	2,196,247	2,098,062	− 1.4	+ 4.7
Net Underwriting Gain (Loss)	(15,912,894)	(24,793,539)	(21,476,947)	− 35.8	+ 15.4
Net Investment Income	21,924,445	19,507,869	17,659,729	+ 12.4	+ 10.5
Operating Earnings after Taxes	6,610,383	(3,327,453)	(2,150,139)	− 298.6	+ 54.8
Policyholders' Surplus	94,288,390	75,511,417	63,809,497	+ 24.9	+ 18.3
Premium to Surplus Ratio	1.87 to 1	1.91 to 1	1.85 to 1		
Dividends to Stockholders	2,806,818	2,723,108	2,342,572	+ 3.1	+ 16.2

(Courtesy: Best's Aggregates & Averages, A.M. Best Co.)

Figure A

part of that formula. Knowing how to write accurate repair estimates and having the skill to perform those repairs are other vital elements in the formula.

Perhaps the most important action the contractor can take is to *form liaisons with insurance adjusters*. A good working relationship with an adjuster will ensure that you get called for repairs, and that you have your name included on the bank draft. Without a good rapport and the assistance of an adjuster, you will not succeed. Finally, what produces success is *action*. Go out and do it.

Chapter 1

UNDERSTANDING INSURANCE COMPANIES, ADJUSTERS AND POLICYHOLDERS

Insurance repair projects involve three parties with distinct interests: the insurance company, the adjuster, and the policyholder. A contractor entering the business of insurance repair should understand the perspective of each of these entities.

The **policyholder** has a financial interest in a property. This individual or entity purchases insurance for the purpose of preventing or minimizing financial loss in the event the property is destroyed. If the property is destroyed, the policyholder wants to receive the maximum possible compensation.

The **insurance company** sells a policy on a property, hoping there will never be a loss. If, in fact, a claim is made, the insurance company prefers to pay the minimum compensation that will reasonably cover the damage.

The **claims adjuster** must weigh the needs of both parties and determine the appropriate compensation for the loss. Although the adjuster is sent by the insurance company to determine the minimum payable amount, the settlement must also be satisfactory to the policyholder.

Figure 1.1 illustrates the relationship between the three parties. The insurance company and the policyholder are the two forces (or weights) on the scale. The claims adjuster is the fulcrum, striving to balance the scale between the two other influencing parties.

The Insurance Company's Role

An insurance company is like a gambler in that it risks certain funds on the possibility it will collect more than it pays out. An insurance company earns its money by "betting" that a group of policyholders' losses (plus its own operational expenses) will not exceed the premiums paid by that same group of policyholders.

For example, the ABC Insurance Company sells homeowner's insurance policies to 1,000 homeowners. The premiums collected for a period of one year total $400,000 ($400 for each policy). The insurance company is counting on the probability that the cumulative losses and expenses for that group of policyholders will not exceed $400,000. If the losses and expenses do amount to less than the insurance company's "break-even" point (premiums minus claims, minus overhead expenses), then it has made a profit (see Figure 1.2).

Anticipating and Processing Claims

Insurance companies calculate probabilities using what is called *Actuarial Science*. With these calculations, they come very close to predicting what their claims payments will be. Since the insurance companies can anticipate a certain percentage of losses or claims, they engage an appropriate number of claims personnel to process those claims.

Some companies, such as Allstate Insurance Company and State Farm Mutual Insurance Company, hire or staff their own claims departments. Others retain the services of an outside claims service company, such as the National Claims Service or Underwriters Adjusting Company. Examples of these two arrangements are given in the following section.

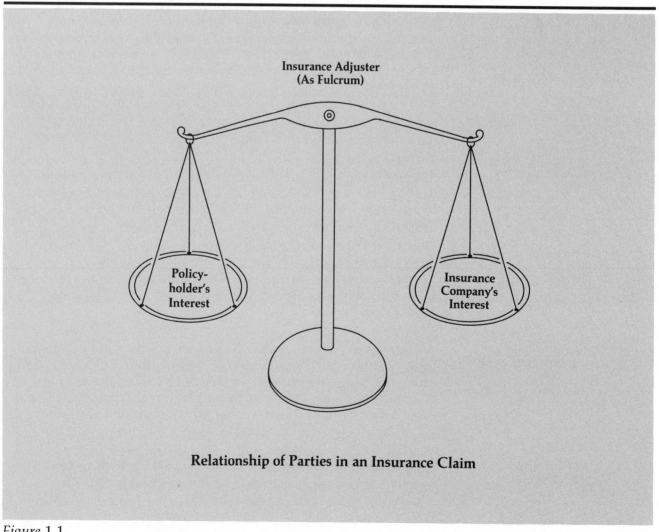

Relationship of Parties in an Insurance Claim

Figure 1.1

Example #1: Property owner, Joe Gotz, buys an insurance policy with State Farm Mutual. When a fire destroys his building, he calls his insurance agent (the person from whom Joe bought the policy) to report the loss. The agent informs the State Farm Claims Service Center, which dispatches an adjuster to investigate the claim. This adjuster is an employee of State Farm. The adjuster then verifies effective dates of the policy, determines whether damages are the result of a peril covered by the policy, and evaluates the damages in order to settle the claim. The adjuster often carries claim checks that he can make payable to customers like Mr. Gotz in settlement of claims that can be established, appraised, and validated quickly.

Example #2: A nightclub owner in Atlanta is having difficulty finding an insurance company to insure his building. A local sales agent introduces the nightclub owner to an insurance broker in Chicago. The broker is able to place the coverage with Peoria Mutual, a specialty insurance company based in Illinois.

When the night club is destroyed by fire, the Chicago broker notifies Peoria Mutual. Peoria Mutual contacts an independent adjuster in Atlanta to investigate and settle the claim, acting as Peoria Mutual's local representative. The independent adjuster is paid an hourly fee for his service. This adjuster may not have settlement checks "in hand" but, if in close contact with Peoria Mutual's home office, can get the claim settled and paid quickly.

In Example #1, State Farm's adjuster is an employee on their payroll. The independent adjuster in Example #2 is an employee of the independent adjusting company, and is paid an hourly rate plus expenses by the hiring insurance company, in this case, Peoria Mutual.

As was previously established, insured losses resulting in claims payments are the determinant of the insurance company's profitability. Another major factor is the expense of hiring claims personnel. The insurance company's solvency depends on minimizing not only the claim payments, but the cost or expense of processing those claims (the adjusting personnel).

From this information, one might conclude that an insurance company would want to fight every claim and do so with the least possible adjusting expense. This is not true. Although the insurance company does not want to pay unwarranted claims, it

Figure 1.2

ABC Insurance Company	
Number of policies sold	1,000
Premium amount of each policy	$ × 400
Total premiums	400,000
Total claims paid	−340,000
Operating Expenses	− 30,000
Profit	**$ 30,000**

cannot afford, nor is there any reason to challenge, every single claim as if it were fraudulent. The insurance company wants legitimate claims processed promptly, while keeping processing expenses (claims or adjusting personnel) to a minimum.

Considerations for the Contractor

If a contractor generates problems on a repair job, the claims personnel must become more involved in the case. This increased involvement boosts the insurance company's claims processing expenses, and naturally discourages the insurance company from hiring that contractor for future work. Contractors interested in pursuing insurance work should have a clear understanding of their role in these projects.

The following is a ridiculous but true story of a problem created by a contractor who actually believed he was doing the insurance company a favor.

The policyholder's home was so untidy that when a small kitchen fire spread soot through the home, it was difficult to tell what was soot, and what was preexisting dirt. The young adjuster made the correct decision to allow for cleaning and painting of walls in the affected area. The painting contractor then decided that the insurance company was being cheated by paying for the already dirty wall to be painted (even though the wall was also lightly covered with soot). He insulted the policyholder, pointing out that the claim was abusive, and encouraged the adjuster to reverse his original decision.

A policyholder complaint to claims management resulted in a meeting between the contractor, claims manager, claims supervisor, and the original adjuster at the home of the policyholder. Four people had to reinspect the loss on behalf of the insurance company, only to conclude that the original decision to paint was "in order."

The claims manager decided that a jury would rule that, because of the soot, the wall would have required painting regardless of the preexisting condition. Five people had to waste half a day due to the inappropriate action of the repair contractor.

This is an example of a costly incident growing out of a very small matter, the painting of a wall. Unfortunately, the same scenario has occurred on a multi-million dollar scale, often because a contractor became involved in an area that did not concern him.

It is never a good idea for contractors to argue with policyholders, nor to try passing themselves off as the insurance company's representatives. If the policyholder is dissatisfied with something that the contractor does or says, and sees the contractor as representing the insurance company, he is likely to take his complaint to the Insurance Commissioner's office. Insurance companies are naturally sensitive about policyholder complaints to the Insurance Commissioner, and may not want to hire the contractor for any future work.

Contractors overpricing or defrauding insurance companies represent another source of higher loss or claim payments. Overpricing may occur when charges are made for restoration work that will not be performed, or when items in an estimate are "padded." For example, a contractor may charge for insulation in a given wall, but never install it. By the time the insurance adjuster returns to inspect the job, the repairs are completed and the insulation, or lack thereof, cannot be detected.

Insurance repair is quite different from new construction in that the structure is not exposed for very long in the course of the work, and inspections are infrequent. As a result, deception is easier to carry out in insurance repair work, and may occur more often. To offset such losses, some insurance companies monitor contractor pricing and actual repairs. The following example is a true case history.

> An adjuster was called upon to reinspect an apartment building after repairs had been made by a contractor to the tune of $250,000. (The original adjustment was handled by another firm.) Upon inspection, the adjuster found that 80% of the charges listed for tearing out and replacing drywall were unfounded. Rather than replace the drywall as allowed for by the original adjuster, the contractor simply cleaned and painted, assuming that the drywall would not be inspected, or that it would be difficult to detect that this cleaned and freshly painted drywall was not new.

> The cost of the contractor's fraud on this repair job was about $30,000, but the cost to his own business was immeasurable. Not only was he never called on again by this particular insurance company, but his reputation suffered and his insurance repair business folded.

Contractors interested in establishing a successful business – whether insurance repair or new construction – must clearly adhere to legitimate pricing methods and honest estimating procedures.

The Adjuster's Role

To understand the process of insurance repair, one must understand the role of the claims adjuster. When the adjuster becomes involved in a case, a whole series of events has already transpired. First, the customer bought the insurance policy out of prudent foresight, having a financial interest in the property and wanting to be protected in the event of a loss.

While the consumer hopes never to need the insurance coverage, he may have been impressed by an advertisement or by an insurance salesperson and decided to purchase the policy. Nearly every step in procuring the insurance will have been a positive, practical move generating a sense of security, trust, and control . . . . until a loss occurs.

When a loss occurs, the policyholder, who may be somewhat distraught, naturally reasons that he has signed the application for insurance coverage, and dutifully paid the premium notices, and that it is now time for the insurance company to live up to its end of the contract. This positive expectation can change quickly because any deficiency in coverage will surface when a loss occurs.

For example, the policy may have been purchased five years earlier based on values that were realistic at that time, but totally inadequate today. Or, the insurer may deny coverage due to a specific exclusion. Coverage would also be denied if there is any finding of "suspicious origin." Several situations could impose restrictive limits on the insurance company's liability, any of which might make for an angry policyholder.

Figure 1.3 is a sample list of exclusions extracted from a standard homeowner's policy. These statements are typical exclusions in this type of policy, and show the wide range of possible

Typical Insurance Company Policy Exclusions

Losses Not Covered:

a) theft committed by an insured person;

b) theft in or from a dwelling under construction, or of materials and supplies for use in construction, until the dwelling is completed and occupied;

c) theft of any property while at any other residence owned, rented to, or occupied by, an insured person, unless the insured person is temporarily residing there:

d) theft of trailers, campers, watercraft, including furnishings, equipment, and outboard motors, away from the residence premises;

e) theft from any part of a residence premises rented by you to other than an insured person.

Exclusions:

We do not cover loss or damage to the property described in the Personal Property Protection coverage resulting directly or indirectly from:

1. Water damage, meaning:

   a) flood, surface water, waves, tidal water or overflow of any body of water, or spray from any of these, whether or not driven by wind;

   b) water which backs up through sewers or drains; or

   c) water below the surface of the ground. This includes water which exerts pressure on, or flows, seeps or leaks through, any part of a building, or other structure, sidewalk, driveway, or swimming pool.

Direct loss that follows water damage and is caused by fire, explosion or theft is covered.

2. Earthquake or other earth movement. Direct loss that follows an earthquake and is caused by fire, explosion or theft is covered.

3. Enforcement of any ordinance or law regulating the construction, repair or demolition of buildings. We will cover loss caused by actions of civil authority to prevent the spread of fire, unless the fire is caused by a loss we do not cover.

4. Neglect of an insured person to take all reasonable steps to save and preserve property at and after a loss, or when the property is endangered by a loss we cover.

5. Nuclear action, meaning nuclear reaction, discharge, radiation or radioactive contamination, or any consequence of any of these. Loss caused by nuclear action is not considered loss by fire, explosion or smoke. Direct loss by fire resulting from nuclear action is covered.

6. War or warlike acts, including insurrection, rebellion, or revolution.

7. Any loss occurring while the hazard is increased by any means within the control or knowledge of an insured person.

An excerpt from a typical insurance company policy showing a sample of common exclusions (found in many policies) which may restrict payment of a claim.

Figure 1.3

limitations that can restrict the payment of restoration claims. Figure 1.4, shows the additional areas whereby restrictions or limitations may affect the amount of the claim payment.

This background of events lead to the insurance company's response to the policyholder's loss and subsequent filing of a claim. The insurance company must send a representative to analyze and resolve the claim. Who does the insurance company send into this potential hornet's nest? Not the ad man from the Madison Avenue agency that created the impressive advertising, nor the salesperson whose advice led to the purchase of the policy, not even the underwriter who wrote the policy. The person sent to analyze and resolve the claim is the **claims adjuster**. The adjuster is often perceived by the public as the "pit bull dog" of the insurance industry, who tries to avoid relinquishing the insurance company's funds. It must be remembered, however, that the adjuster is also the person who signs the settlement check. The adjuster needs cooperation to expedite the job for all involved parties.

Responsibilities of the Adjuster

The claims adjuster (or claims representative) has three basic functions in handling any given claim:

1. **To establish coverage**, that is, to determine whether there is, in fact, insurance coverage in force, and whether the terms of the policy have been complied with by the policyholder.
2. **To determine legal liability**, that is, whether the insurance company is legally obligated to pay.
3. **To determine damages**, or the amount that should be paid. An adjuster depends heavily on the services of a contractor in carrying out this third function.

Let us consider the first function, that of **verifying coverage**. When an adjuster is called upon to assess a claim, he or she follows a few simple procedures to determine whether coverage is in order. First, the adjuster validates the policy number and checks to see if the loss being claimed occurred within the effective dates of the policy. The next step is to find out what happened, and whether those events meet the criteria of the policy's coverage. Is the occurrence that resulted in the loss actually described in the policy as a covered peril? The adjuster then determines whether there will be any limiting factors, such as a coinsurance penalty, depreciation, or deductible applied to any settlement. Once it has been established that there are no applicable exclusions nor any breach of contract by the policyholder, then coverage is in order and effect.

The second function of the adjuster is to determine the legal **liability** of the insurance company to the policyholder. This is a simple matter inasmuch as the insurance policy wording obligates the insurer to the extent that coverage is in order and effect.

The third function of an adjuster is to determine the **damages**, or how much should be paid. This is the area where adjusters depend heavily on general contractors. It is here where a contractor's competence, honesty, and reliability can make him invaluable to an insurance company adjuster.

Additional Limitations in Coverage

1. We will pay only when a covered loss exceeds the deductible shown on the declarations page, and then only for the excess amount, unless we have otherwise indicated in this policy.

2. No more than one deductible shall apply to loss by windstorm or hail arising out of one occurrence.

3. In the event of a loss, we will not pay for more than the insurable interest that an insured person has in the property covered by this policy, nor more than the amount of coverage afforded by this policy.

4. A covered property loss will be settled on an actual cash value basis. This means there may be a deduction for depreciation.

5. In making an actual cash value settlement, payment will not exceed the smallest of the following amounts:

   a) the actual cash value at the time of the loss;

   b) the amount necessary to repair or replace the damaged property; or

   c) the limit of liability applying to the property.

6. We will settle any covered loss with you. We will pay you unless another payee is named in the policy. We will pay within 60 days after the amount of loss is finally determined. This amount may be determined by an agreement between you and us, a court judgment, or an appraisal award.

7. If you and we fail to agree on the amount of loss, either party may make written demand for an appraisal. Each party will select a competent and disinterested appraiser and notify the other of the appraiser's identity within 20 days after the demand is received. The appraisers will select a competent and impartial umpire. If the appraisers are unable to agree upon an umpire within 15 days, you or we can ask a judge of a court of record in the state where the residence premises is located to select an umpire.

   a) The appraisers shall then determine the amount of loss, stating separately the actual cash value and the amount of loss to each item. If the appraisers submit a written report of an agreement to us, the amount agreed upon shall be the amount of loss. If they cannot agree, they will submit their differences to the umpire. A written award by any two will determine the amount of loss.

   b) Each party will pay the appraiser it chooses, and equally bear expenses for the umpire and all other expenses of the appraisal.

Figure 1.4

Identifying the Damage

One might think that establishing damages in an insurance claim is fairly simple and clear-cut. After all, property is tangible; it is either damaged or it is not. Unfortunately, this can be a very complicated aspect of the insurance repair business. The following example illustrates the potential for complications.

The lower level of a two-story dwelling has been flooded with water, leaking through the ceiling from a malfunctioning plumbing fixture in the upper level. The homeowner has called his/her own contractor who immediately condemned the drywall in both walls and ceiling. The adjuster knows that there is a possibility that the walls may not even have been exposed to the water. Nevertheless, the homeowner has been told that once drywall gets wet, it will turn to paste and eventually fall off. The homeowner insists on replacing all of the drywall.

An insurance repair contractor can help the adjuster settle this dispute in one of the following ways:

1. *Offering his own second expert opinion as to the soundness of the drywall.*

2. *Having a patch cut out of the wall (at insurance company expense) to check the interior of the drywall for moisture.*

It should be noted that while the insurance repair contractor may not always agree with the adjuster, his sound advice, based on experience and evidence, provides the rationale required by the adjuster to reverse his decision. The next example illustrates this point.

A substantial loss occurred to an apartment building. The owner called in his own contractor, who wrote a repair estimate for $380,000. The adjuster on the case could not come to an agreement with that contractor on the repairability of various systems in the structure. The adjuster maintained that the plumbing, electrical, and HVAC systems on several floors did not warrant replacement, and came up with an estimate of $290,000.

The adjuster then called in a reputable and respected contractor and asked for his opinion. This contractor agreed with the owner's contractor. After reinspecting and some additional testing (performed by the insurance company's repair contractor), the adjuster realized that he was going to have to make some concessions. While it was determined that the owner's contractor had overestimated some labor time, the adjuster was going to have to allow for some additional mechanical work. A compromise was reached, and the loss was settled for $329,000.

This example shows the situations that may arise in the process of trying to establish the damages. Other possibilities might be a salvage dispute over wet carpeting (can it be saved?), or the repairability of an oak floor that has been submerged in water. Was lightning the cause of an air conditioning failure or was it just an old compressor?

The insurance repair contractor, no matter what his specialty, is in a position to serve as an expert on behalf of a claims adjuster. The adjuster needs the contractor's judgment, skills, experience, and an honest opinion.

Assigning Costs

The adjuster must establish not only the extent and type of damage, but also a fair legitimate cost for the particular repair. Part of the service provided by the contractor is providing an appropriate cost estimate. Most adjusters are trained and experienced in estimate preparation, but writing an estimate that no contractor will agree to is useless and will never get a claim settled. For this reason, adjusters normally try to have an estimate reviewed and agreed to by a contractor. The approved estimate is referred to, in the insurance industry, as an *agreed price estimate*. The preparation of repair estimates and the role of the adjuster and contractor in reaching an agreed price will be covered in Chapter 4. It is, however, important to note here, that the method most adjusters use is *unit price estimating*.

Unit Price Estimating: A unit price list is developed by establishing the labor and material cost for each item of repair or replacement per a specific unit of measure. For example, to install one square foot (the unit of measure) of fiberglass, foil-faced batt insulation (R11, 3.5″ x 15″), the unit price as of January 1, 1989, is $.38. This is a base (bare) cost. If overhead and profit are to be included (installed cost), the unit cost is $.47 per S.F. Based on this unit cost, the total cost for 1,000 square feet of insulation would be $380. Since the unit cost in this case includes both labor and material, that $380 represents the "installed" price. The unit price used in this example was taken from the 1989 edition of *Means Repair & Remodeling Cost Data* (see Figure 1.5). This and other cost reference sources are discussed in more detail in Chapter 4.

If an adjuster and a contractor have previously agreed to use the same source of unit prices, then the process of reaching an agreed price estimate will be that much easier. This is the preferred method of many adjusters. If the adjuster knows that he can get a particular repair item accomplished for a certain price per square foot (as in the case of the insulation in the previous example), then he need only be sure to accurately perform the quantity takeoff.

The unit price method, dividing a specific repair operation into units of measure, to which costs for labor and material are designated, can be applied to any type of work to be performed on a construction site. A categorized list of items for unit price cost estimating is shown in Figure 1.6.

Adjusters prefer to operate in the field using a unit price list. Since very few insurance companies develop their own unit price list for use by their adjusters, contractors can make themselves more marketable for insurance repair work by developing and maintaining their own list. Some contractors develop such a list, put it on their letterhead, or have it made into a small booklet with their name or logo on it. They then distribute these unit price guides to adjusters as a marketing tool.

Means Unit Price Estimating is a source of direct advice for preparing better unit cost estimates. This guide, together with the Means construction cost data references, should enable the estimator (contractor) to put together a practical price list. Such a list will not only be appreciated by claims adjusters, but will promote their use of the contractor's services. Unit price estimating will be covered in greater detail in Chapter 4 of this book.

072 100 | Building Insulation

			CREW	DAILY OUTPUT	MAN-HOURS	UNIT	BARE COSTS				TOTAL INCL O&P	
							MAT.	LABOR	EQUIP.	TOTAL		
118	0010	**WALL OR CEILING INSUL., NON-RIGID**										118
	0040	Fiberglass, kraft faced, batts or blankets										
	0060	3-½" thick, R11, 11" wide ⑨⓪	1 Carp	1,150	.007	S.F.	.22	.15		.37	.48	
	0080	15" wide		1,600	.005		.22	.11		.33	.41	
	0100	23" wide		1,600	.005		.25	.11		.36	.45	
	0140	6" thick, R19, 11" wide		1,000	.008		.36	.17		.53	.67	
	0160	15" wide		1,350	.006		.36	.13		.49	.60	
	0180	23" wide		1,600	.005		.36	.11		.47	.57	
	0200	9" thick, R30, 15" wide		1,150	.007		.50	.15		.65	.79	
	0220	23" wide		1,350	.006		.50	.13		.63	.75	
	0240	12" thick, R38, 15" wide		1,000	.008		.72	.17		.89	1.07	
	0260	23" wide		1,350	.006		.72	.13		.85	1	
	0400	Fiberglass, foil faced, batts or blankets										
	0420	3-½" thick, R11, 15" wide	1 Carp	1,600	.005	S.F.	.27	.11		.38	.47	
	0440	23" wide		1,600	.005		.27	.11		.38	.47	
	0460	6" thick, R19, 15" wide		1,350	.006		.39	.13		.52	.63	
	0480	23" wide		1,600	.005		.39	.11		.50	.60	
	0500	9" thick, R30, 15" wide		1,150	.007		.56	.15		.71	.86	
	0550	23" wide		1,350	.006		.56	.13		.69	.82	
	0800	Fiberglass, unfaced, batts or blankets										
	0820	3-½" thick, R11, 15" wide	1 Carp	1,350	.006	S.F.	.20	.13		.33	.42	
	0830	23" wide		1,600	.005		.20	.11		.31	.39	
	0860	6" thick, R19, 15" wide		1,150	.007		.33	.15		.48	.60	
	0880	23" wide		1,350	.006		.33	.13		.46	.57	
	0900	9" thick, R30, 15" wide		1,000	.008		.54	.17		.71	.87	
	0920	23" wide		1,150	.007		.54	.15		.69	.83	
	0940	12" thick, R38, 15" wide		1,000	.008		.67	.17		.84	1.01	
	0960	23" wide		1,150	.007		.67	.15		.82	.98	
	1300	Mineral fiber batts, kraft faced										
	1320	3-½" thick, R13	1 Carp	1,600	.005	S.F.	.28	.11		.39	.48	
	1340	6" thick, R19		1,600	.005		.44	.11		.55	.66	
	1380	10" thick, R30		1,350	.006		.75	.13		.88	1.03	
	1900	For foil backing, add					.06			.06	.07	
	9000	Minimum labor/equipment charge	1 Carp	4	2	Job		43		43	69	

072 200 | Roof & Deck Insulation

			CREW	DAILY OUTPUT	MAN-HOURS	UNIT	BARE COSTS				TOTAL INCL O&P	
							MAT.	LABOR	EQUIP.	TOTAL		
203	0010	**ROOF DECK INSULATION**										203
	0030	Fiberboard, mineral, 1" thick, R2.78	1 Rofc	800	.010	S.F.	.27	.20		.47	.64	
	0080	1-½" thick, R4		800	.010		.44	.20		.64	.82	
	0100	2" thick, R5.26		800	.010		.61	.20		.81	1.01	
	0300	Fiberglass, in 3' x 4' or 4' x 8' sheets										
	0400	15/16" thick, R3.3	1 Rofc	1,000	.008	S.F.	.39	.16		.55	.70	
	0460	1-1/16" thick, R3.8		1,000	.008		.50	.16		.66	.82	
	0600	1-5/16" thick, R5.3		1,000	.008		.62	.16		.78	.95	
	0650	1-5/8" thick, R5.7		1,000	.008		.73	.16		.89	1.08	
	0700	1-7/8" thick, R7.7		1,000	.008		.78	.16		.94	1.13	
	0800	2-¼" thick, R8		800	.010		.78	.20		.98	1.20	
	0900	Fiberglass and urethane composite, 3' x 4' sheets										
	1000	1-11/16" thick, R11.1	1 Rofc	1,000	.008	S.F.	.62	.16		.78	.95	
	1200	2" thick, R14.3		800	.010		.73	.20		.93	1.14	
	1300	2-5/8" thick, R18.2		800	.010		.96	.20		1.16	1.40	
	1500	Foamglass, 2' x 4' sheets, rectangular										
	1510	1-½" thick R3.95	1 Rofc	800	.010	S.F.	1.63	.20		1.83	2.13	
	1520	2" thick R5.26		800	.010		2.06	.20		2.26	2.61	
	1530	3" thick R7.89		700	.011		2.44	.23		2.67	3.07	
	1540	4" thick R10.53		700	.011		4.33	.23		4.56	5.15	
	1600	Tapered 1/16", 1/8" or ¼" per foot				B.F.	1.10			1.10	1.21	
	1650	Perlite, 2' x 4' sheets										

Figure 1.5

UNIT PRICE SECTION

TABLE OF CONTENTS

(from *Means Repair & Remodeling Cost Data, 1989*)

Figure 1.6

Additional Considerations for the Contractor

In addition to identifying pricing, and possibly repairing the insured damage, the successful insurance repair contractor must perform in a professional and competent manner. If the contractor is called in by the adjuster, the policyholder sees the contractor as under the supervision and authority of the adjuster. For example, if the contractor does not show up when scheduled to do the work, the policyholder blames the adjuster. If the contractor's employees misbehave, the adjuster tends to be held responsible by the policyholder. Contractors who want repeat business sent their way by an adjuster should take care not to cause trouble for the adjuster due to any lack of professionalism or competence. The following example illustrates the sort of negative repercussions that can be avoided with a little care and foresight.

> A policyholder sustained a fire loss. The adjuster arranged to meet the contractor at the policyholder's home to assess the damage. The insurance purchased for the home had been chosen on the basis of the company's service-oriented advertising and financial strength. The adjuster arrived at the home a little early in order to review the claims process with the policyholder and put her at ease. They waited in the front yard for the contractor.

> The contractor arrived late, and with a crew of two workmen who were unclean, unshaven, and ill-mannered. Their vehicle was dirty, dented, rusty, and barely running. Not surprisingly, the policyholder began to doubt the adjuster's judgment and credibility. Furthermore, the contractor did not get the job. Although the adjuster knew and trusted the repairman's abilities, he could not convince the homeowner to contract his services.

While appearances may be deceiving, and the laborers may, in fact, be honest, hard workers, the lead contractor should simply have left them in the van and handled this meeting alone.

As far as the van goes, construction equipment and vehicles can naturally get dirty and banged up. Nevertheless, it is a good idea to clean up the business vehicle for a meeting with a potential client. A dented and rusted vehicle might make the homeowner feel that the contractor is not financially stable, or unable to handle the repairs.

Finally, it is important to be on time. The property owner is considering whether he will use this contractor to repair the damages. The basis for this decision rests, in part, on whether the contractor will report to the job site on time and on schedule, and whether or not he will finish the job on time and on schedule. The homeowner may reason that if the contractor is unable to show up on time for the first meeting (which is most important because it may win him the job), then perhaps he is not the right contractor for this repair job. Furthermore, the homeowner's time is valuable, the adjuster's time is valuable, and the contractor's livelihood is at stake.

Having Adequate Resources

Performing in a professional and competent manner also means that a contractor should not accept a restoration job unless he has the resources to perform this work. This means having the available trade skills and funds. The following examples illustrate the problems that may arise as a result of inadequate resources.

The XXX contracting company had devoted its business efforts to expanding its work in insurance repair. The company succeeded in developing relationships with adjusters so that, at one point, it had seven repair jobs underway simultaneously. Due to the size of the contracting company, it had reached its maximum capacity. That is, it could not handle an additional job effectively.

When the company was called by three different adjusters to bid and possibly undertake three additional jobs, XXX proceeded to bid and actually was awarded two of the projects, despite the fact that the company lacked the resources to handle these jobs. The firm divided their total work commitments among their available tradesmen and supervisors. Given too little individual attention, none of the projects progressed rapidly enough to satisfy the property owners or the insurance companies. This slow progress was a negative reflection on the competence of the contracting firm. One of the projects was actually taken away from XXX (while in progress) and given to another contractor. This change caused additional disruption and expense to the insurance company and XXX was dropped from their approved list of contractors.

Another problem may arise when a contractor takes on a repair job that he cannot afford. While it is true that many of the claims are paid up front, a contractor does not want to put himself in the position of having to wait for the insurance adjuster's door to open in the morning in order to pick up the check to finance that day's repairs. For example:

A small, two-man restoration firm took on a $20,000 repair job. The coverage was in order and effect so there was no question that the insurance company would provide the funds. The adjuster and the contractor had already agreed on the estimate, so they knew how much was going to be paid. All the necessary paperwork had been done so as to ensure the contractor's legal rights to proper compensation. The adjuster ordered issuance of the settlement check, but it was understood that the check would not be ready for seven days. Even then, it would be in the form of a bank draft, so that it would take ten days to clear after it was deposited. The adjuster promised he had included the contractor's name on the draft and requested that the repairmen begin the job. They would have liked to begin, but three-to-four thousand dollars worth of material was needed to start the job. Granted, this is a small sum to many contractors, but it was more than this small operation could afford to finance. The adjuster saw that this job was out of their league, and lost confidence in the credibility of the firm.

The same kind of problem could arise on a multi-million dollar repair job. A larger contracting firm should consider the same issues.

Insurance repair jobs come in all sizes. Contractors must understand their capabilities and their limits. Adjusters appreciate an honest, straightforward approach, and are more likely to call when they are confident that a certain contractor is capable of doing a job of a particular size.

The Policyholder's Role

A policyholder is someone who has given prudent consideration to the protection of his assets. These include his home or business establishment. The policyholder buys insurance to "cover" himself financially in the event of a loss. When a loss does occur, the policyholder tends not to expect to have to pay anything out of his pocket. Any suggestion to the contrary may be met with resistance.

The policyholder who has suffered a casualty needs to be made "whole" again. However, he often overlooks the insurance company's interpretation of what being made "whole" means. The insurance company asserts that being made "whole" means being restored to "pre-loss" condition, however good or bad that might have been. For example:

> A home, while functionally sound, may have been in need of an exterior paint job. The paint, applied 20 years earlier, was cracked and peeling when a hail storm not only damaged the paint further, but also damaged the siding to the extent that a new paint job was needed.

> The insurance adjuster inspecting the dwelling decides that the policy does indeed cover the paint job. However, he feels that the insurance company should pay only 50 percent of the cost, based on the following logic. The policy promises only to make the policyholder's covered property "whole." That is, the property should be restored to its "pre-loss" condition. Clearly, the adjuster is not going to find a workman to recreate the splitting, cracking, and peeling condition of the pre-loss 20-year-old paint job. On the other hand, if he allows for a completely new paint job, he is paying to restore the siding to a much better condition than its pre-loss state. Allowing for a completely new paint job without any deduction for depreciation conflicts with the principle of insurance and indemnity.

> The only way to balance or "adjust" the scales while repairing the damage is to have the house painted, but to pay only a portion of the cost (50 percent). The policyholder was overdue for a paint job anyway. As a result, he will only have to pay 50 percent of the total cost that he would normally have paid. Of course, a policyholder may argue that he is now "worse off" because he has to spend 50 percent of the cost of a paint job, whereas he would otherwise have been perfectly happy with the cracked and peeling paint. In such cases, the policyholder may pocket the 50 percent issued by the insurance company and go another 20 years without painting.

Factors that Limit Coverage

There are several limiting factors in insurance coverage that affect how much a homeowner will collect. In some situations there is simply no coverage. For example, a new electrical code may require an upgrade in electrical service. Insurance policies usually do not cover such costs incurred in order to meet current code requirements.

In another circumstance, the policyholder who does not purchase enough insurance may be penalized. Or, there may be a deductible which is subtracted from the claim payment. Deductibles may run as high as $500 on dwellings and $10,000 on commercial buildings. Restrictions on payment may also occur in the form of depreciation or "betterment," as in the example of the home with cracked and peeling paint.

Many policyholders feel that after having paid insurance premiums for many years, and having sustained the aggravation and/or trauma of the loss in question, they do not deserve to have their claim payment reduced by any of these additional limiting factors. In fact, some policyholders have gone to great lengths to avoid spending any of their own money due to the application of these limiting factors. Most often, it is the innocent contractor who suffers in these cases. The problem arises when the contractor has completed repairs. He may have received 70% of his fee from the insurance company, but the remainder (not paid by the insurance company due to limiting factors) is still outstanding. Supposedly, the remaining 30 percent was to be paid by the homeowner, who now admits he does not have the money.

The contractor who is creative and flexible can avoid these problems by addressing the situation up front, before the job begins. Chapter 5 covers such arrangements in more detail. The following examples show how a contractor might approach a homeowner, at the outset, in order to avoid problems in collecting payment.

Example #1: The insurance company agrees that a homeowner has sustained a loss of $8,000 and, after applying the $100 deductible, issues a check for $7,900. The contractor comes to an agreement with the homeowner that if he is awarded the job, he will waive the collection of the deductible amount. The contractor is happy because he gets the job. The homeowner is happy because he did not have to pay any of his own money.

Example #2: The insurance company agrees that a store owner has sustained a loss of $77,000. However, there is a "coinsurance" penalty, and the claim payment is reduced by 20 percent, or $15,400. The claim is further reduced because $1,500 of the repair bid represents electrical work required due to an improved code. (Increased repair cost required to bring the property up to an improved building code is not usually covered by insurance policies.) Finally, there is a $1,000 deductible. The final payment to the insured is $59,100 (see Figure 1.7).

The contractor bidding the job might make an arrangement with the insured, or policyholder, to perform the major structural and mechanical repairs. They agree that the tearing out and removal of debris, cleaning and painting, and other work requiring unskilled labor are to be done by the insured at his own expense. The insured decides to use his existing personnel to perform these less technical tasks over an extended period of time. With this part of the work eliminated, the contractor estimates that he can do the remaining work for $59,100. The insured is able to manage the insurance reductions by absorbing the cost into his existing payroll, using his own staff. Again, everyone is happy (see Figure 1.8).

Effect of Coinsurance Penalty on Insurance Premium	
Agreed Repair	$77,000
Less Coinsurance Penalty	−15,400
Less Cost of Code Improvements	− 1,500
Less Deductible	− 1,000
Insurance Pays	$59,100

Figure 1.7

What would have happened if the contractor had been inflexible? He might have demanded the total amount of $77,000 (including the $17,900 which is not covered by insurance) up front before repairs were started. Since the insured either did not have this amount or did not want to commit this amount of cash, the contractor would have lost the opportunity. He probably would not have gotten the job. The insured would have collected the $59,100 and gone out to find someone who would do the major repairs within his budget.

There are other kinds of arrangements that contractors can make with policyholders in order to obtain an insurance repair job. It is important to know the insured's needs and to think in terms of meeting those needs. The possibilities might include negotiating a slight change in the floor plan, or some other minor adjustment.

One must be careful, however, that these "deals" remain ethical and in no way defraud the insurance company. Nor should a contractor be overly aggressive and offer "deals" that are not invited or necessitated by the limiting factors in insurance coverage. *Let the policyholder express his needs.*

The Contractor's Performance

Finally, a policyholder needs the contractor's commitment to do the specified work competently and efficiently with minimum "disruption" to his or her life.

One successful contractor has some good ideas on how to keep the property owner happy. When he is awarded a repair job, he prepares a schedule, or itinerary. He lets the property owner know when his crew will start; which days they will be there, and what they will be doing. He also lets them know when he plans to be finished. Figure 1.9 shows a project schedule (taken from *Means Forms for Building Construction Professionals*) that can be used for this purpose. Because the calendar scale is blank, it can be used for days, weeks, or months. Since the job description is blank, the form can be used to fit any repair situation.

Such information (the project schedule) must be compiled anyway to be used internally by a construction company, but is also a great benefit to a property owner, who may be given a more basic version. If, for some reason, the contractor or one of his workmen

Contractor's Proposed Solution	
Insured would owe this amount, not covered by insurance	$17,900
Insured performs work, valued at $17,900, using his/her own staff	−17,900
Insured owes repair contractor	$ 0

Figure 1.8

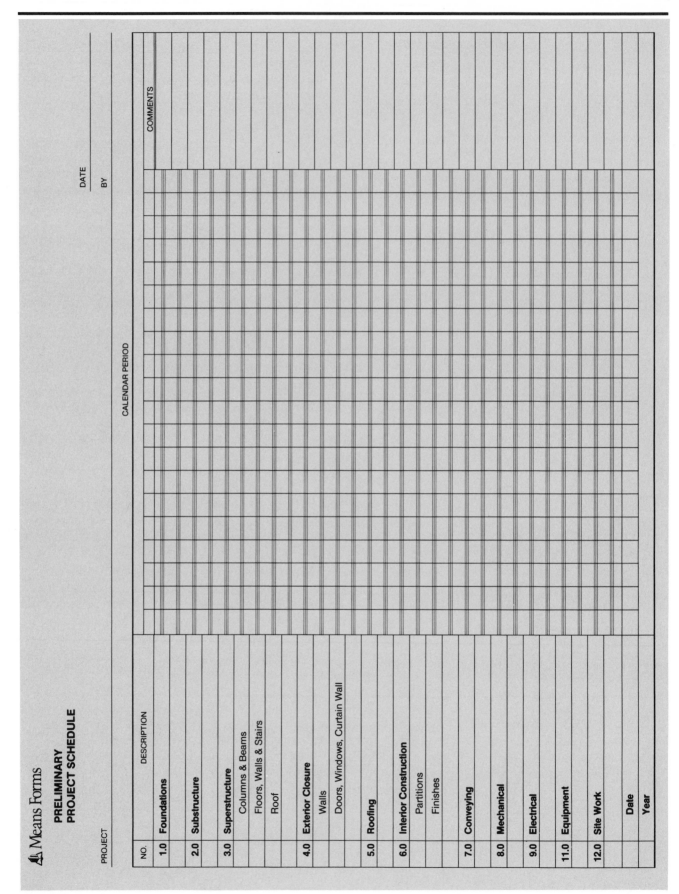

Figure 1.9

18

cannot be there at the specified time(s), the property owner is informed. Good communication prevents problems.

The contractor does not promise what he cannot deliver. He allows himself some leeway. For example, if he believes a particular job will take 14 days, he may tell the property owner that it will take 16 days. The extra two days will cover any problem or snag in scheduling. It is far better to under-promise and over-deliver than to over-promise and under-deliver.

Summary

In summary, the following is a list of basic principles that the contractor should remember in seeking a profitable relationship with insurance adjusters and policyholders.

- Use fair pricing of labor and material.
- Legitimately estimate damages.
- Show competence in work performance.
- Use disciplined employees who conduct themselves in a professional manner.
- Maintain a professional image in all dealings with insurance company representatives and the policyholder.
- Earn the policyholder's confidence with professionalism, both in the initial inspection of the site, and in carrying out the work.
- Be flexible to meet the needs of the policyholder.
- Do not try to act as a representative of the insurance company.
- Do not create problems that generate complaints.
- Do not try to explain insurance coverage or any limiting factors to the policyholder.
- Do not try to win a repair job by being overly aggressive, offering uninvited "deals," or other such tactics.
- Do not allow the members of the crew to intimidate or be disrespectful to policyholders.

Chapter 2
THE INSURANCE REPAIR CONTRACTOR

The Importance of a Professional Image

What does a contractor or subcontractor achieve, and how does appearance and the way he conducts himself fit into this plan? Why is it so important to act and look completely professional?

To answer these questions, it helps to consider the needs of insurance adjusters and policyholders. First of all, despite the fact that the contractor called by the insurance company does not officially represent the insurance company, the property owner will tend to associate these two entities. Whatever the contractor does or says, and however he may appear, it reflects on the insurance adjuster who has called him in. Consequently, if the contractor speaks, acts, or appears in a negative way, problems are caused for the adjuster, who will be less inclined to recall this contractor for future work.

Secondly, the property owner is looking for someone he can trust to restore his damaged property. Whether the damage involves a minor plumbing repair or a major fire repair, the contractor should look as if he knows exactly what he is doing. The contractor's meeting with the property owner is crucial in creating a good first impression; the contractor who appears disorganized or unprofessional does not give a favorable impression and probably will not get the job.

The contractor's attitude and actions should be both respectful and professional. Communications should be up-front and honest. If there is a controversial area of damage, the contractor might raise the issue and initiate a discussion between the adjuster and policyholder. If the damages concern an area outside of the contractor's expertise, he should inform the adjuster that a subcontractor should be called in to analyze the situation. For example, a general contractor may be confronted with the issue of whether an air conditioner condensing unit has been damaged internally. If he is not a skilled HVAC mechanic, he should not try to speculate. Such a decision should be deferred to a certified HVAC tradesman.

The Resume
It goes without saying that contractors must be competent and experienced in their particular specialty. If not, they should seek additional knowledge and experience. Contractors pursuing

insurance work should prepare a brief resume on their training and experience, including some references from people for whom they have worked. Include information such as birthplace, military record, and special community involvement. Such items can elicit a positive reaction from both homeowners and claims adjusters who may share common experience or interests with the contractor. Figure 2.1 provides an idea of the sort of information some insurance repair contractors include on a typical resume.

Jobs have been awarded because a property owner had been in the navy and the contractor was a former Sea-Bee. Likewise, one contractor was hired because he had listed in his resume that he supported the Special Olympics. He had been called upon yearly by the group and always made some donation. The property owner happened to have some connection and made a decision based on the contractor's support of the Special Olympics. Once the contractor's competence is established, these little connections can be the deciding factor. The property owners, in this case, felt that this was a man who could be entrusted with the repair of their dwelling.

Appearance

With regard to physical appearance, a contractor should be neat, clean, and dressed appropriately for his particular field. Clearly, a carpenter does not need to wear a suit. On the other hand, he should not wear a dirty tee shirt with ragged blue jeans.

Appearance and Conduct of Personnel

Some contractors establish—formally or informally—a set of "Rules of Conduct" for their employees or subcontractors to follow while on the property owner's premises. This can be a very important procedure since the standards of conduct on an insurance repair job should be quite different from those that are acceptable on a new construction site.

Sample Resume

Ben Good, General Contractor

- 15 years experience remodeling and insurance restoration
- Unit price estimates
- Client list available upon request
- Supporter, Atlanta Lions Club
- Credit & bank references
- Licensed, State of Georgia #780721608
- Bonded, U.S. Fidelity & Guaranty Co.
- Insured, Lexington Insurance Co. # 818746

Figure 2.1

On a new construction site, workers are among their peers. A property owner is rarely present. Under those circumstances, behavior standards might not have to be highly structured. By contrast, during repair and restoration work, a property owner is almost always present. The work is being done on a structure that may serve as the policyholder's dwelling or business base. This in itself is a potentially stressful situation. Furthermore, the event that led to the damage, or loss, may have been traumatic, and the property owner may be impatient with what he considers a "loose" approach to the work. "Rowdy" or unprofessional behavior may result in complaints by the owner to the claims adjuster – the contractor's customer. Figure 2.2 is an example of what an insurance repair contractor's "Rules of Conduct" might include.

Contractors' Offices and Warehouses

Some larger insurance repair contractors maintain warehouses for storing construction materials, equipment, and vehicles. Usually these structures also house the contractor's administrative offices, which are furnished to accommodate visiting adjusters.

Many years ago, a South Florida contractor frequently invited adjusters to come in and use his telephones and office facilities when they needed a place to do paperwork, away from their own offices. He would provide them with technical information, clerical help, and almost anything else they thought they needed. He kept a well-stocked liquor cabinet and a refrigerator filled with sandwiches and snacks. In today's legal and insurance climate, providing such extensive benefits would be considered outlandish and inappropriate, and we do not recommend it. Nevertheless, there will be times when an adjuster or property owner has to

---

**Sample Rules of Conduct for Employees
of the Insurance Repair Contractor**

- No profanity.
- No loud noise (except for construction equipment). No loud music, talking, or shouting.
- Minimal communication with property owner (communicate with job superintendent for owner-related issues).
- No rowdiness or "roughhousing."
- Do not enter areas of building where you have no business.
- Refer all problems or questions to job superintendent.
- When job superintendent is communicating with property owner or adjuster, afford them privacy.
- Do not touch or draw attention to the building owner's personal property.

Figure 2.2

visit the contractor's facilities for legitimate reasons. No matter how humble the facilities, they should be clean and reflect an organized, professional attitude.

Condition of Vehicles

By the nature of the job, construction vehicles can get dirty and dented easily. Nevertheless, before meeting an insurance claims representative or a property owner, it is a good idea for contractors to clean their vehicles. Although such impressions may be ill-founded, a dirty, dented, and rusted vehicle might make the property owner question the contractor's financial stability or ability to handle the repair job. Owners may also wonder whether the vehicle's shabby condition is a reflection of the kind of workmanship the contractor would perform on their property.

Licensing, Bonding, and Contractors' Liability

A property owner needs to know that the contractor bidding the work on his property is capable of doing the job. He may know very little about the contractor. What kind of training and experience does the contractor have? Does the contractor operate his business professionally and honorably? Does he have the financial resources to start and complete the job? If something goes wrong, will he make good on his commitment?

A property owner could spend a lot of time and energy attempting to find answers to those questions. Fortunately, he/she can obtain this information by simply requesting the contractor's licensing and bonding criteria.

When a contractor is licensed by a municipality, it shows that he is committed to his trade, at least to the extent that he has registered with the authorities. In some areas, licensing requires certain training or minimum skill standards verifiable by written, oral, and even performance testing. By requesting the contractor's license number, the property owner assures himself that the contractor has specific capabilities.

As a further guarantee that a job will be completed, the contractor purchases a bond that assures the property owner that the work will be performed, even if the contractor defaults on his contract agreement. This might be a Performance Bond, a Completion Bond, or a Fidelity Bond.

Licensing
The following true accounts illustrate the importance of licensing and bonding.

A policyholder was suspected of deliberately destroying—by fire—his commercial establishment. One clue to potential fraud was his claim for obviously inflated building damage. (This is a common pattern in fraudulent claims: suspicious origin and inflated damages.) The policyholder hired an attorney who then located a general contractor willing to submit an inflated bid for the amount of $510,000 for restoration.

The adjuster called in a reputable insurance repair contractor, and the two men undertook to establish a fair and realistic repair figure. Their estimate came to $320,000. Meanwhile, the adjuster's investigation did not result in enough incriminating evidence to document the policyholder's involvement in the deliberate destruction of the insured property. Consequently, it looked as if the insurance *company would have to honor the claim. Now the big issue was how much they were going to pay: the inflated, fraudulent amount or the legitimate figure. The adjuster planned to stick by the legitimate estimate and apply limiting factors (depreciation, coinsurance, and the deductible) to the greatest extent possible. He attempted to offer a settlement based on the $320,000, less limiting factors. The policyholder rejected this offer and filed a lawsuit against the insurance company on the grounds that the damages were not fairly computed by the adjuster.*

The first step in defending the lawsuit was for the insurance company to support their position with expert witnesses and documentation. Their key expert was the insurance repair contractor. At the first opportunity, the plaintiff's attorney (policyholder's attorney) cross-examined the insurance repair contractor. It was quickly determined that the insurance repair contractor, although reputable and competent, had failed to secure proper licensing. A year or two before, he had considered it, but concluded that the license was of little use to him in securing work. Furthermore, it might subject him to new government regulations with which he disagreed. Operating without a license was considered a misdemeanor and carried a very small fine. He simply thought that he would never need it. Now he was in a courtroom testifying on behalf of a valuable insurance client, and he was being discredited for not being properly licensed. The plaintiff's attorney pressed the issue until the judge agreed to consider the inadmissability of the insurance repair contractor's testimony. The jury was now questioning the insurance repair contractor's credibility instead of focusing on the real issue: the inflated claim. As a result, the insurance company's position was compromised, and the insurance repair contractor had seriously damaged his relationship with this company.

Contractors should find out what their municipalities require in order to secure proper licensing in a particular field. County, city, or state construction authorities should be consulted for information regarding licensing. Contractors are advised to comply with local stipulations in obtaining a license, and to print the license number on business cards and stationery.

Bonding

The second account deals with the issue of bonding and the kinds of problems that may arise when proper bonds have not been obtained.

A shopping center sustained a substantial fire loss. The insurance adjuster knew he had a loss of over two million dollars. Very few insurance repair contractors are available in a given state who can handle the financial demands of such a large loss. Furthermore, few large "new construction" contractors are interested in insurance repair work. The policyholder expressed a preference for a particular general contractor. The contractor was financially sound, and had been in business for 20 years. He prepared an estimate for $3.2 million.

The adjuster then called in an insurance repair contractor who, with the adjuster's assistance, prepared an estimate of $2.4 million. A third estimate was also requested, using a third contractor who was paid to prepare an "independent" figure, which he calculated to be $2.8 million.

The policyholder's attorney was reasonable but astute, expressing an interest in resolving the matter as equitably as possible. After reviewing and analyzing the figures submitted by all, he concluded that the main differences were in certain labor and management costs, along with profit margins. The attorney admitted he had no argument with the insurance repair contractor's figure, and that upon presenting proof of a Performance Bond, the repairs could begin.

Unfortunately, construction bonds are not easy to procure, and in this case the insurance repair contractor was not able to procure one. Therefore, his estimate was of no use to the adjuster or the policyholder.

The adjuster had no legal right to force the policyholder to use an unbondable contractor, and if he had tried to force the use of this contractor, he might have left his company exposed to future lawsuits should anything go wrong. The claim was finally settled for $3.2 million.

Most insurance repair work does not require construction bonds. However, contractors providing estimates for large jobs should make sure that the adjuster is aware, in advance, of their capacity to secure a bond.

Fidelity Bonds

Fidelity Bonds, which protect against employee dishonesty, are much easier to secure than construction bonds. In most cases, when a small contractor indicates that he is "bonded," he is usually referring to a Fidelity Bond. Having the word "bonded" imprinted on stationery and business cards makes a good impression, and is highly recommended.

Contractors' Liability Insurance

Contractors' Liability Insurance is an area often misunderstood by contractors, policyholders, and some adjusters. To understand the purpose of this form of insurance and its benefit to a potential customer, let us ask why a customer would want the contractor to carry Contractors' Liability Insurance. The property owner may think that it protects him if the contractor does inferior work. However, this is not the purpose of this type of insurance. No Contractors' Liability policy will cover the contractor (and consequently, the property owner's funds) for bad workmanship. Generally speaking, a Contractors' Liability policy covers the contractor against damages for which he may become legally liable, resulting from his own or his employee's negligence. The Contractor's Liability policy excludes the *specific item* or *object* on which he is working. The underwriters who originally wrote such policies intended to protect the contractor against liability in general, but they reasoned that the contractor should be induced to exercise the utmost level of care in the area in which he himself is most proficient. Therefore, for example, the expert in hanging chandeliers should never drop one. If he does, his insurer will refuse to pay for it.

Furthermore, if the insurer promised a warranty on workmanship, then anyone could perform perfunctory or inferior work and the insurer would pick up the tab. By refusing to cover workmanship and property in the "care, custody, and control" of the contractor, the insurer makes it the contractor's responsibility to be proficient and careful with his own work items. Liability insurance covers injury to persons or property that are *outside* of the contractor's area of control.

The following examples and illustrations demonstrate liability insurance coverage.

Example #1: An electrician is hired by the general contractor to install several light fixtures in an expensive dwelling. One fixture is an $18,000 chandelier purchased separately by the homeowner for installation by the electrician. In the process of lifting the chandelier to the vaulted ceiling, the electrician drops it to the ground and it is destroyed. Even though the contractor is legally liable to the property owner, there would be no coverage for this loss under his liability policy due to the fact that the chandelier was under the "care, custody, and control" of the electrician (and, therefore, the contractor).

Example #2: An interior paneling installer whose work involves only one small area of a building stops for a cigarette break. He inadvertently tosses a lit match into a bucket containing a highly flammable adhesive. The resulting fire destroys the entire building. This loss would be covered by the contractor's liability insurance because it is outside of his area of "care, custody, and control." On the other hand, if only the paneling were damaged, the insurer would pay nothing.

Example #3: A pest control operator erects a fumigation tent to enclose an entire two-story building. On his way out of the building, just before the insecticide gas is about to be introduced, he inadvertently leaves a high intensity bulb burning in an upstairs room. Upon contact with the heated bulb, the gas ignites and fire destroys the building. There would be no coverage under a Contractors' Liability policy due to the "care, custody, and control" exclusion.

The insurer demands that the contractor exercise utmost care in his field of expertise. That is why the "care, custody, and control" exclusion was written into the Contractors' Liability policy. The contractor in this example would likely be legally liable for the destroyed building, but he would be technically uninsured for that incident. If the burning building caused damage to neighboring property, the neighboring property would be covered. This is because the neighboring property was not in the "care, custody, and control" of the fumigation contractor (the insured).

There are no insurance policies which cover a contractor's workmanship, or property in his "care, custody, and control."

Despite the fact that Contractors' Liability insurance is not all-inclusive, it is still recommended. Keep in mind that it is worth shopping around to find the best rates. Do not purchase an excess of liability insurance. Consult with an accountant, attorney, or insurance agent to determine the right amount. Contractors who do carry Contractors' Liability insurance should let potential customers, including insurance company representatives, know that they have this coverage by listing it on advertisements or on business cards.

Advertising Contracting Services

Since insurance coverage is written on virtually everything, tangible or intangible, it is safe to say that just about every service or skill is needed by an insurance company at one time or another. It is important for contractors to clearly state the particular service(s) they provide. Contractors can use that specialty to introduce themselves to the insurance industry. It may be a mechanical trade, such as plumbing or electrical work, or it may even be a sub-specialty within one of those trades.

Examples

The following examples illustrate ways to obtain insurance repair jobs for specialty contracting.

Plumbers

A certain number of plumbers can make a handsome living installing bathroom tub enclosures alone. A common insurance claim involves leaking shower or tub enclosures. The homeowner notices water escaping from the enclosure. In some cases, water escapes over an extended period of time before it is noticed. It may result in staining of floor coverings or, in the case of an upstairs enclosure, can even cause the collapse of a ceiling below. Usually, when this type of loss is covered, the restoration involves tear-out of the existing enclosure and reinstallation of a new shower or tub system. This procedure also requires wall and, possibly, floor tile work and carpentry.

Plumbers specializing in this particular type of work might prepare a flyer and send it to adjusters in their region. The flyer should describe the service that they provide. (See Figure 2.3 for

Claims for Leaking Tub Enclosures??

We INSPECT and REPAIR TUB ENCLOSURES and other plumbing-related losses for Insurance Companies.

- WE ARE CERTIFIED PLUMBERS
- LICENSED, BONDED, AND INSURED
- WE CHECK FOR CAUSE & ORIGIN OF LEAK
- WE REPORT ORALLY OR IN WRITING, AS REQUESTED
- WE APPRAISE RESTORATION USING UNIT PRICING

Honest John, Plumber
(803) 555-8644

Figure 2.3

a sample advertisement.) These specialists might then be called on for any claim situation involving a tub/shower enclosure, to determine the cause of the leak, and to make a proposal on the repairs. This might be a service that no one else is providing in a particular location. Because the plumbers use a standard unit price for these enclosures, the adjusters will always know how much it is going to cost to have the work done. This simplifies the adjuster's job and encourages him to continue using the same contractor.

Air Conditioning Mechanics

Air conditioning mechanics have applied the same approach, specializing in checking lightning damage to air conditioners. This type of damage represents another common claim in some regions of the country. Adjusters use guidelines established by engineers to determine whether there is enough evidence to attribute the failure of a compressor to lightning. The problem is that some of the guidelines pose questions that only a journeyman could test for an answer.

Adjusters are likely to call on this type of subcontractor for such an analysis because he knows exactly what information the adjusters require to make a decision. Furthermore, this air conditioning specialist has established prices that can be applied to every inspection. Consequently, adjusters always know how much he will charge. Figure 2.4 is a sample advertisement for inspection and repair of lightning damage to air conditioners.

General Contractors

Some subcontractors acquire skills in restoration work as it relates to their trade, and then spread word of their experience among general contractors who specialize in insurance repair. These subcontractors see the general contractor who specializes in insurance repair work as their source of income.

Contractors whose specialty is general contracting (restoration) should identify themselves as such. Some general contractors specialize in a certain project size, or dollar classification. One Atlanta contractor solicits only those insurance loss jobs that are under $10,000. His preference is to have twenty $10,000 jobs, rather than one $200,000 job. This decision may be a reflection of his management style, or it may be that his operation generates more profit on smaller jobs than on larger ones for such work. Another general contractor may prefer to deal only with insurance losses that exceed $500,000, and he may be willing to travel anywhere in the country. Another contractor from Texas may specialize in extinguishing oil rig fires . . . anywhere in the world.

How Much to Charge

No matter what service the contractor offers to the insurance industry, it is important to establish unit prices for those services. This pricing must be done up-front so that an adjuster knows how much the contractor will charge the insurance company for almost any job in that specialty. The tub enclosure specialists mentioned in the previous section might charge $1,150 to tear out and reinstall a shower pan and complete tub enclosure. This might be their standard price for this job, including three tiled walls. Occasionally, they might encounter unforeseen problems and lose some money, but they should make a good profit most of the time. They are providing a reliable service at a predictable price, and should, consequently, obtain a good deal of regular work. The adjusters who call them always know up-front what a job is going to cost.

Many insurance adjusters have learned, by experience, to be skeptical, and may be inclined to mistrust contractors. Preparing and providing a unit price list (in advance) for the services offered can go a long way towards allaying their suspicions and doubts.

Means Repair & Remodeling Cost Data and *Means Open Shop Building Construction Cost Data* are annually-updated books that serve as comprehensive resources for unit price estimating on commercial and residential remodeling. The former is based on union labor rates, while the latter uses "open shop," or non-union, labor rates.

Means Repair & Remodeling Cost Data can be used to verify an estimate, to find the right cost data for unusual products and difficult installations, and to estimate labor costs. It contains a

Figure 2.4

detailed analysis of labor productivity, material costs, labor rates, and equipment costs. Figure 2.5 is a page from *Repair and Remodeling Cost Data* showing the subjects that are included, and the extent to which each is covered.

In many regions of the country, open shop labor rates prevail. This factor makes it more difficult for a contractor to establish unit prices since, even within one region, labor rates vary. A reference such as *Means Open Shop Building Construction Cost Data* can be a great advantage in such cases. Means open shop prices have already been compiled covering many components of building construction.

HOW TO USE
THIS BOOK

HOW THE BOOK IS ARRANGED

This book is divided into four sections: Unit Price, Assemblies, Reference and an Appendix.

Unit Price Section: All cost information has been divided into the 16 C.S.I. divisions. These divisions are patterned after the MASTERFORMAT created and adopted by the Construction Specifications Institute, Inc. This system is widely used by most segments of the building construction industry.

A listing of all divisions and an outline of their subdivisions is shown in the Table of Contents page at the beginning of the Unit Price Section.

Each unit price line item has been assigned a 10 digit code. A graphic explanation of the numbering system is shown on the "How to Use Unit Price" page.

Descriptions Each line item number is followed by a description of the item. Sub-items and additional sizes are indented beneath appropriate line items. The first line or two of the main (bold face) item often contain descriptive information that pertains to all line items beneath this bold face listing.

Crew The first column after the description lists the typical crew needed to install the item. Typical crews are defined and listed at the beginning of the Unit Price Section. When an installation is done by one trade and requires no power equipment, the appropriate trade is listed in the crew column. For example, "2 Carp" indicates that the installation is done with 2 carpenters. If, however, the installing crew is listed as "C-2", it is made up of 1 carpenter foreman, 4 carpenters, 1 building laborer, plus an allowance for power tools. The crew costs are listed both with bare labor rates, and with the installing contractor's overhead and profit. For each, the total cost per eight-hour day and the composite hourly cost for the crew are listed.

Crew Equipment Cost The power equipment required for each crew is included in the crew cost. The daily cost for crew equipment is based on dividing the weekly bare rental rate by 5 (number of working days per week), and then adding the hourly operating cost times 8 (hours per day). This "Crew Equipment Cost" is listed in Division 016.

Daily Output To the right of every "Crew" code listing, a "Daily Output" figure is given. This is the number of units that the listed crew will install in a normal 8-hour day.

Man-hours The column following "Daily Output" is "Man-hours". This figure represents the man-hours required to install one "unit" of work. Unit man-hours are calculated by dividing the total daily crew hours (as seen in the Crew Tables) by the Daily Output.

Unit To the right of the "Man-hour" column is the "Unit" column. The abbreviated designations indicate the unit upon which the price, production and crew are based. See the Appendix for a complete list of abbreviations.

Material The first column under the "Bare Cost" heading lists the unit material cost for the line item. This figure is the "bare" material cost with no overhead and profit allowances included. This cost is developed by contacting manufacturers, dealers, distributors and contractors throughout the United States. Prices shown reflect the national average of these quoted prices for January of the current year.

Labor The second column under the "Bare Cost" heading is the unit labor cost. This cost is derived by dividing the daily labor cost by the daily output. The wage rates used are 30 city union averages and are listed on the inside back cover.

Equipment The third column under the "Bare Cost" heading lists the unit equipment cost. This figure is the daily crew equipment cost divided by the daily output.

Total The last column under the "Bare Cost" heading lists the total bare cost of the item. This is the arithmetic total of the three previous columns: "Material", "Labor", and "Equipment".

Total Incl. O&P The figure in this column is the sum of three components: the bare material cost plus 10%; the bare labor cost plus overhead and profit (per the billing rate table on the inside back cover); and the bare equipment cost plus 10%. A sample calculation is shown on the "How to Use Unit Price" page preceding the Unit Price Section.

Assemblies Section: This section uses an "Assemblies" format that groups all the functional elements of a building into 12 "Uniformat" Construction Divisions.

At the top of each "Assembly" cost table is an illustration, a brief description, and the design criteria used to develop the cost. Each of the components and its contributing cost to the system is shown. For a complete breakdown and explanation of a typical "Assemblies" page, see "How to Use Assemblies Cost Tables" at the beginning of this section.

Material These cost figures include a standard 10% markup for "handling". They are national average material costs as of January of the current year and include delivery to the jobsite.

Installation The installation costs include labor and equipment, plus a markup for the installing contractor's overhead and profit.

Reference Section: Following the items in the "Unit Price" pages, there are frequently found large numbers in circles. These numbers refer the reader to information and data in this Reference Section. This material includes estimating procedures, alternate pricing methods, technical data and cost derivations. These derivations illustrate the development of costs and may be of value in listing materials for purchase. This section also includes information on design and economy in construction.

Appendix: Included in this section are Historical and City Cost Indexes, a Project Schedule (CPM) example, a list of abbreviations and a comprehensive index.

Historical Cost Index This index provides data to adjust construction costs over time.

City Cost Indexes These indexes provide data to adjust the "national average" costs in this book to 162 major cities throughout the U.S. and Canada.

CPM (Critical Path Method) A brief review of project scheduling is included in the Appendix. By following the example, it can be seen that even a small project can benefit from CPM scheduling. Realistic duration times for activities are developed using the crews and man-hours shown in the line items of the manual.

Abbreviations/Index A listing of the abbreviations used throughout this book, along with the terms they represent is included. Following the abbreviations list is an index for all sections.

Figure 2.5

Chapter 3

GETTING THE ATTENTION OF CLAIMS ADJUSTERS

Many professional contractors or other construction tradesmen find the idea of selling their services distasteful. Nevertheless, it is important to develop sales skills. If the most competent, conscientious, and cost efficient repairman or contractor cannot sell his services to a property owner or claims adjuster, his building skills will never bring him a profit.

This is not to say that a contractor should practice "hard sell" techniques, pestering adjusters to call him and trying to push his way into jobs. Aggressive sales people often try to sell something we do not want or need, and cannot afford. They make a science of learning how to overcome a prospect's objections, even though the objections may be well-founded. "Hard sell" salesmen refuse to accept the word "no" from a prospect and persist to the point of being ridiculous or offensive. This unpleasant picture of sales is held by many people. It is, however, completely inappropriate in the insurance repair business. High pressure sales would never be welcomed by insurance adjusters, who tend to be skeptical and would mistrust a contractor who used such methods.

Once the contractor has recognized the need to market his services, the next step is to find out who are the insurance industry representatives who handle claims involving damaged property and how to contact them. What can contractors do to ensure that the adjuster calls the next time a claim arises involving damaged property? How can contractors develop a working relationship with an adjuster so that they are called again? This chapter presents effective methods contractors can use to market their services to insurance adjusters without using negative sales pressure.

Independent Versus Exclusive Agencies

When a policyholder experiences a loss, unless instructed otherwise, the first person he calls is his insurance agent (the one who sold him the policy). This agent may be independent – that is, self-employed and representing several different insurance companies. Or, the agent may be an employee – an exclusive agent of one specific insurance company.

An example of the independent agency is the Vidaillet Insurance Agency representing *The Hartford Insurance Group, American Hardware Mutual, Aetna Life & Casualty,* and *Continental Insurance Company.* The sign in front of this agent's office might appear as shown in Figure 3.1.

An example of the exclusive agent would be J. R. Crosa, agent for *Eagle State* Insurance Company. This agent's storefront or sign might look like that shown in Figure 3.2.

Independent Agents

Most independent insurance sales agencies have one person on their staff designated as their claims clerk. This clerk records the information from the policyholder, and then notifies the particular insurance company that a loss has occurred. If the situation warrants, the claims clerk might call in a contractor on an emergency basis even before an adjuster is assigned. The claims clerk decides which contractor to call; the contractor seeking insurance repair work must make his services known to this person. The following example illustrates this point:

A policyholder calls his independent agent because he has just discovered that his house has been flooded, and is standing in two feet of water. After obtaining the necessary information from the policyholder, the agency claims clerk calls the insurance company to report the incident. To prevent further damages, the claims clerk calls a contractor (perhaps one specializing in carpet restoration, including water extraction) to extract the water immediately. Some contractors

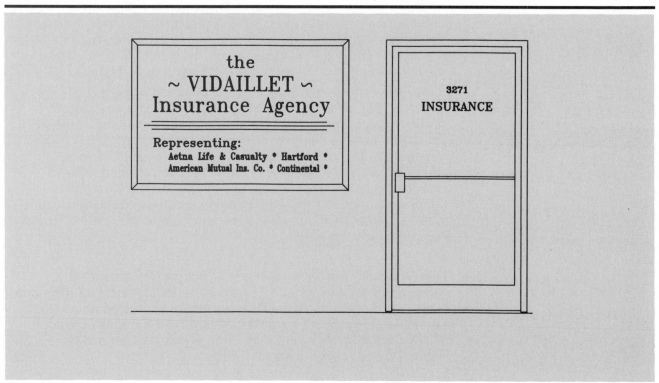

Figure 3.1

are available on an around-the-clock basis for just such an emergency; the claims clerk will know the names of such contractors.

Figure 3.3 is a schematic diagram showing the path followed when a loss is reported through the independent agency system.

Exclusive Agents

In contrast, the exclusive agent does not usually record the information on a claim from the policyholder, but rather refers the policyholder directly to their parent company's claims department, which is better suited to taking the initial reports than the exclusive agent. In fact, policyholders are often advised of this reporting procedure before a loss occurs. The exclusive agency reporting process is illustrated in Figure 3.4, and further explained using the following example.

Again, using the example of the policyholder's house flooded with two feet of water, let us assume that he purchased his policy from an exclusive agent who instructed him to call the claims department directly in the event of a loss. When the loss occurred, he called the claims office. The claims representative took down the information, created a claim file and immediately called in a contractor for emergency water extraction.

Contractors who are equipped and willing to perform services on an emergency, around-the-clock basis should make themselves known to the company claims representative.

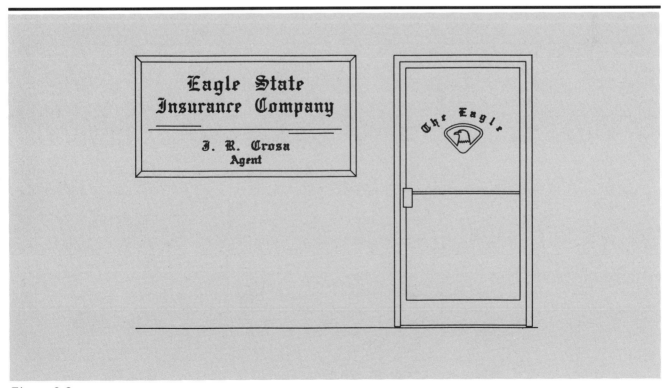

Figure 3.2

Contacting the Right Person

In summary, the contractor wants to make contact with the claims clerk at the independent insurance agencies or the claims departments of the parent organizations that hire exclusive agents. The idea is to find out who has the authority to choose and call in contractors, and then to establish contact with that person.

In some cases, exclusive agents do handle small claims. For example, the State Farm Insurance Company provides most of its agents with the authority to settle claims up to specified limits. That is, they can settle smaller claims themselves without having to contact the claims department. Therefore, it may be worth calling the offices of this type of exclusive agent, to find out who is responsible for handling small claims in that particular agent's office.

Making the First Contact

The first step in establishing contacts is to make a list of all the independent insurance agencies in the contractor's operating territory. Next, call each one to find out and record the names of their claims clerks. Introduce yourself to the claims clerks and even call on them periodically, about every four to eight weeks. (Conduct for such personal visits will be discussed later in this chapter.)

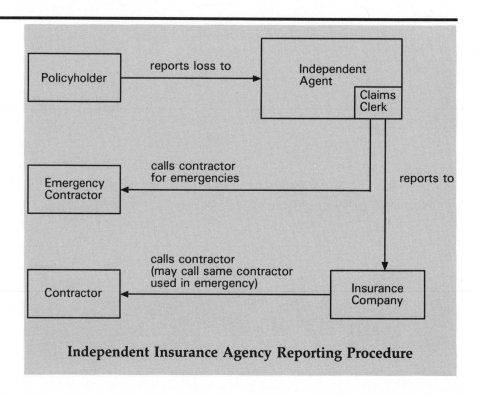

Independent Insurance Agency Reporting Procedure

Figure 3.3

The term *insurance agency* always refers to the person or company selling the insurance, not the parent insurance company or the claims adjuster. The agency includes the commissioned sales agent and his office crew.

Getting back to the question of "where do you find or how do you meet these people?", an excellent place to start is the local Yellow Pages. Another source is the Yellow Pages of the major cities in a state or region, particularly the state capitol (usually available at a local public library). Insurance claims contacts will be listed under "Insurance Agencies," "Insurance Companies," and "Adjusters."

Professional claims associations are another good source of contacts. Their meetings provide an excellent opportunity for making important contacts. Some associations allow repair contractors to become associate members. (Professional claims associations throughout the U.S. are listed in the Appendix.)

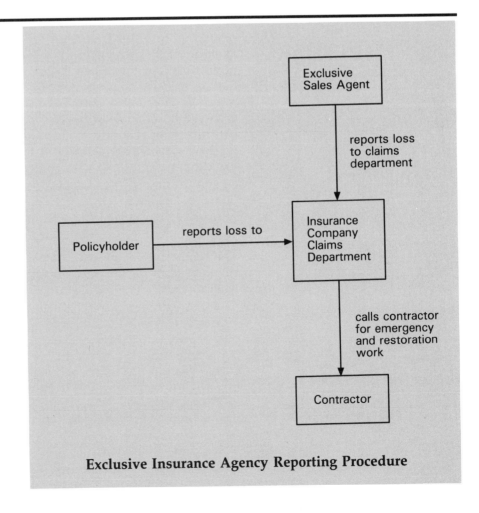

Exclusive Insurance Agency Reporting Procedure

Figure 3.4

Following are sources of information for developing a list of insurance agencies to call:

- The Yellow Pages
- The Bell System Industrial Pages
- The Insurance Commissioner's Office (see listings of state Insurance Commissioners in the Appendix)
- The Professional Insurance Agents Association (National Headquarters: 400 N. Washington St., Alexandria, VA 22314)
- The Independent Insurance Agents of America, Inc. (National Headquarters: 100 Church Street, New York, NY 10007)
- The National Association of Casualty & Surety Agents (National Headquarters: 85 John Street, New York, NY 10038) Local chapters of the above three may be listed in the local White Pages or can be obtained through the area's Insurance Commissioner's office.
- The Underwriter's Handbook (for your state), which is ordered from: The National Underwriter Company, 420 E. Fourth Street, Cincinnati, OH 45202
- National Association of Independent Insurance Adjusters N.A.I.I.A., 222 West Adams, Chicago, IL 60606
- *Claims Magazine* mailing list of subscribers
- *Claims Magazine*, 1001 Fourth Avenue Plaza Suite 3029, Seattle, WA 98154
- The Claims Service Guide. For a copy write to: The Bar List Publishing Co., P. O. Box 948, Northbrook, IL 60065

The Claims Department Hierarchy

Contractors should understand the hierarchy of insurance claims personnel in order to further identify the person who is most likely to call a contractor for a repair job. For example, there is no point in trying to contact the claims manager when the assistant manager is the one who is responsible for calling in a contractor. The following description is of a typical hierarchy, but organizational structures vary from company to company. The contractor must routinely "qualify" claims employees, which means establishing whether or not they have the authority to decide which contractor to call in on a job.

Find out who at each office is on the "front line." That is, who would be the first person to take the initial report from the policyholder? Getting to know that person is the first step toward getting one's "foot in the door."

The hierarchy at the insurance company usually follows this order (illustrated in Figure 3.5), starting with the person who has the least authority:

- Claims Clerk
- Telephone Adjuster, sometimes called a *Telephone Claims Representative* or *TCR*
- Field Adjuster
- Senior Adjuster, sometimes called a *General Adjuster*
- Claims Supervisor, sometimes called the *Claims Examiner* or *Analyst*
- Claims Unit Manager, who oversees a small crew of supervisors
- Claims Manager

There are many variations of this hierarchy, from company to company. A small branch office may only have one Claims Clerk and one Claims Manager. Another company may have the same two positions, but may refer to the Claims Clerk as *Claims Secretary*, and the Claims Manager as *Resident Adjuster*. Figures 3.6 and 3.7 illustrate some typical office structures.

Senior Property Adjuster

Contractors interested in insurance repair should take special note of the Senior Property Adjuster, also known as the *General Adjuster*. These are highly experienced "specialized" adjusters, who handle losses in excess of $25,000 throughout a multi-state region. A General Adjuster might specialize in property claims; general contractors interested in larger loss repair work should contact General Adjusters.

General contractors targeting the restoration work in the $25,000-and-up category will want to identify and contact claims people with titles such as:

- Property Claims Supervisor
- Property General Adjuster
- Property Claims Examiner or Analyst
- Executive General Adjuster

Figure 3.8 is a separate list of key claims personnel whom contractors should contact.

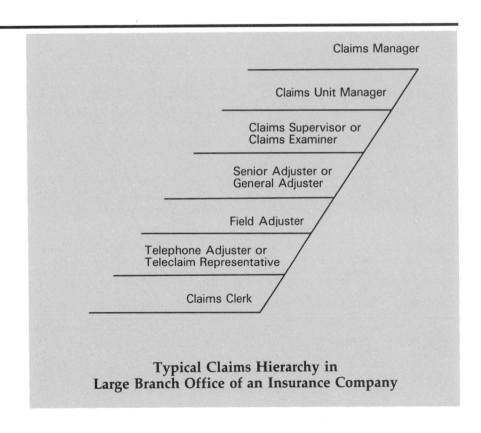

Claims Manager

Claims Unit Manager

Claims Supervisor or
Claims Examiner

Senior Adjuster or
General Adjuster

Field Adjuster

Telephone Adjuster or
Teleclaim Representative

Claims Clerk

**Typical Claims Hierarchy in
Large Branch Office of an Insurance Company**

Figure 3.5

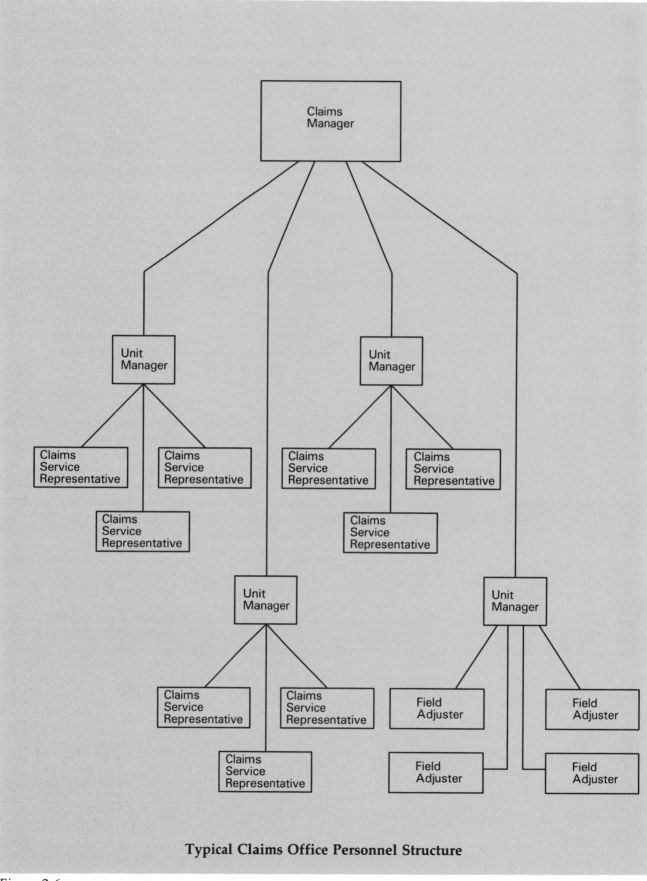

Typical Claims Office Personnel Structure

Figure 3.6

It is, of course, to the contractor's advantage to be known by all of these people. However, for contractors geared to restoration work in the $250 to $10,000 range, it is more important to be known by the lower level claims personnel. When claims are called in, they are received first by the lower level claims people. If there is an emergency situation requiring a contractor's immediate attention, it is the lower level claims person who will make the decision to call in a contractor. Thereafter, it may be a natural step to move up the hierarchy and develop relationships with the next higher level of claims personnel.

Telephone Adjuster

Some insurance companies do not use field adjusters for all claims, but handle minor claims by telephone. The telephone adjuster contacts the policyholder, conducts the necessary investigation, and arranges for a damage assessment by telephone. Telephone adjusters usually require a contractor's estimate. Some contractors find it advantageous to develop a relationship (based on trust) with the telephone adjusters, becoming the "eyes and ears" of the telephone adjuster.

Field Adjuster

The field adjuster is assigned cases that require actual inspection for assessment of damages. Field adjusters are usually trained to prepare their own estimates of building damages. The generally preferred practice is for the adjuster to prepare an estimate while a contractor prepares a separate, independent one. The two then come together to work out differences and arrive at a final agreed

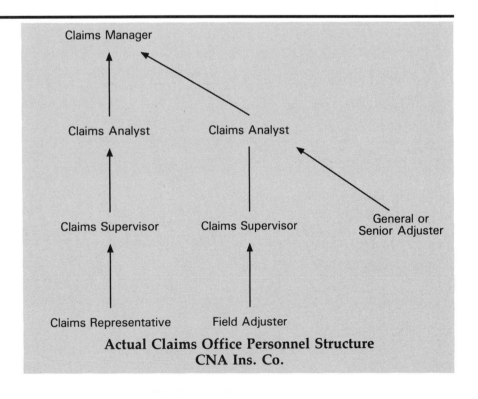

**Actual Claims Office Personnel Structure
CNA Ins. Co.**

Figure 3.7

price. Frequently, an adjuster will bypass this process and simply rely on a chosen contractor to provide him with an estimate. The adjuster then copies that estimate and presents it as his own.

This practice is not condoned by the upper management of claims departments, but it continues "unofficially." The reasons for this situation are many, the most prominent being (1) heavy case loads: the adjuster does not have time to prepare his own estimate on every claim, or (2) inexperience.

Heavy Case Loads

The average "manageable" case load for a field adjuster in a metropolitan area is 40 to 60 claims per month. Given a 21 workday month, that means two to three assignments or cases per day. A telephone claims representative (TCR) who does not do any field work may receive double that figure and still manage his claims adequately. A claims examiner who merely supervises the work done by others may reasonably be expected to oversee 15 to 20 cases per day.

There are some claims departments where field adjusters are assigned upwards of 10 claims per day; TCR's 10 to 15 claims per day, and examiners up to 30 claims per day. Adjusters with such heavy case loads may be tempted to pass off the responsibility of writing estimates to someone else.

Insurance paperwork is viewed by some to be tedious and, at times, boring. This, combined with the freedom of a company car and a flexible schedule, sometimes produces an undisciplined

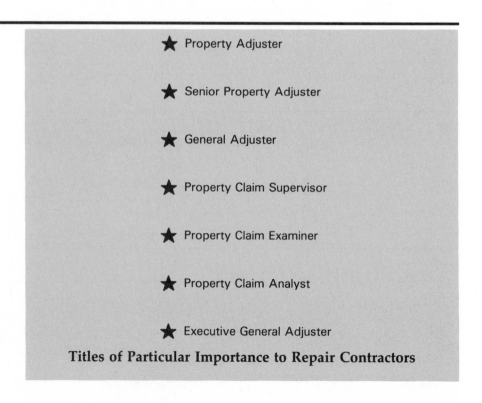

★ Property Adjuster

★ Senior Property Adjuster

★ General Adjuster

★ Property Claim Supervisor

★ Property Claim Examiner

★ Property Claim Analyst

★ Executive General Adjuster

Titles of Particular Importance to Repair Contractors

Figure 3.8

adjuster who cannot seem to find time or the desire to write his own estimates. Of course, this portrait is of a minority of claims people, but keep in mind that occasionally such people are encountered.

Inexperience

As to inexperience; this is a problem usually related to lack of training. There are a few occasions when an adjuster is assigned to a task that he is clearly unprepared to handle. For some companies, a trained and competent staff is not a high priority; they make a minimal investment in personnel. Typically, an adjuster from such a company relies heavily on a contractor for estimate writing.

Another reason for adjuster inexperience may be that he or she "falls out of practice" as a result of having relied on others for too long to prepare their estimates. Again, these situations are the exception rather than the rule.

Working with the Field Adjuster

The contractor should not be concerned with the reasons why an adjuster requests an estimate. Whatever the reasons for the estimate, a contractor interested in insurance work should be willing to prepare one. Remember, the contractor who develops a good relationship with an adjuster will have a steady flow of insurance repair work.

Beware that there are some adjusters who will "use" a contractor for an estimate, not caring whether the contractor actually gets the job. Such adjusters may "use" contractors purely as a source of free estimates, having no consideration for the expenses incurred in this process. It may take a while to identify these adjusters and "filter them out." For example:

A contractor advertises and begins getting calls from adjusters requesting inspection of damaged property and estimates. The contractor starts to win some of these jobs. However, he soon realizes that there is one adjuster who calls frequently and gets estimates, but never hires him to perform the work. This adjuster never meets the contractor at the scene of the damage. He settles his claim and pays the property owner directly. He may even encourage the property owner to use another contractor. The adjuster never gives any indication that he is sympathetic to the contractor's interests, yet he keeps calling for "free estimates."

It would be a mistake to react too quickly in "filtering out" this adjuster. The adjuster may genuinely want to develop a good working relationship, but has simply not had the opportunity to hire the contractor to perform repairs.

The contractor should express his concern to an adjuster who has been calling for "free estimates," but has never awarded him a job, asking directly if he has any suggestions. The contractor should communicate a desire to help the adjuster, while meeting his own obligations. The adjuster's response should reveal his attitude. There is no point in trying to work with an adjuster who could care less for contractors or their financial concerns, yet is ready to accept the benefit of having such a contractor available at a moment's notice when a loss occurs.

An ethical adjuster does not call a contractor to submit a bid unless that contractor has a legitimate chance to actually get the job. If the policyholder already has a favored contractor, the adjuster should prepare an estimate and try to work out any discrepancies with that contractor. If, by chance, the differences cannot be settled and another contractor is needed for an expert opinion, a fair adjuster will make it clear that it is unlikely the second contractor will get this job. Even better, an effort should be made to pay the insurance contractor a fee for submitting the estimate which is, in effect, serving as an expert opinion. In some regions of the country (particularly the Northeast), it is common practice to pay insurance contractors a fee for preparing estimates.

Claims Supervisor or Examiner

Claims supervisors and/or claims examiners oversee the paperwork turned in by the field adjusters. The field adjuster performs the job in the field, and the supervisor or examiner processes the paperwork. (These individuals are sometimes referred to as "paper movers.")

Examiners have usually spent two-to-five years handling claims as field adjusters. Thereafter they supervise or "examine" the work done by field adjusters. Occasionally, a claims examiner has never worked in the field, and one may wonder how this person can supervise a field adjuster and provide the necessary technical direction. Nevertheless, some insurance companies believe that an examiner does not have to be able to *execute* a competent adjustment, simply to *critique* and *approve* it.

For the most part, claims examiners are proficient and experienced adjusters.

Rarely does a claims examiner suggest a contractor. By the time this person reviews a claim, contractors may have already been called. This is also true of claims managers. In fact, most unit managers (who supervise claims examiners) and claims managers (who manage all claims personnel) rarely handle claims.

The positions and the actual work done by the claims people are fairly universal among insurance companies. For example:

> *The XXX Insurance Company has three claims people in their Atlanta office. One is titled* Claims Manager, *and the other two are titled* Claims Supervisors. *However, in the office, all three do the same job supervising outside independent adjusters. As to the differences in their titles, even though each performs the function of claims examiner, one of them is "in charge."*

Independent Adjuster

Some insurance companies do not have field adjusting personnel. The XXX Insurance Company mentioned above is a typical example of such a company. When they receive a claim that requires a field adjuster, they retain the services of an independent adjuster. The independent adjuster acts as if he were the field adjuster employed by XXX Insurance Company. He or she performs the necessary field investigation, assesses damages, and reports back to the claims examiners or, as titled by XXX, the *claims supervisors*.

Independent adjusters are usually paid by the hour. That is, they are paid strictly based on how much time was spent processing a particular claim. Fees vary from claim to claim, because one may be more involved and require more of the adjuster's time.

Do not be mislead into thinking that independent adjusters earn more money if they "cut," or reduce, a claim. The reality is just the opposite. The insurance company that pays the independent adjuster's fee expects to see a small fee if the claim is of low dollar value, and a larger fee if the claim is of high dollar value. A large value claim logically requires more (of the adjuster's) time to prepare than a smaller value claim. This concept can be understood by the following example:

> An adjuster handles a loss for the XXX Insurance Company. The claim settles for $1,250. The adjuster sends his bill for services rendered in the amount of $500. The claims examiner at XXX sees the adjuster's bill and reacts as follows: "What do you mean a $500 adjuster's bill? There's no way you spent that kind of time on this loss; it was only a $1,250 claim!"

> The next time the adjuster handles a loss for this examiner, he may reason (and this is false reasoning, but it does happen) that his service invoice will be compared or measured in terms of the amount of the claim settlement. If he follows this fallacy, he will be inclined to submit a more liberal estimate for the claim in order to "justify" his service bill and prevent it from being challenged.

Independent adjusters may tend to be more inclined to make concessions on controversial items. Ethical independent adjusters should take special care to prevent this attitude among adjusters. On the other hand, this trend in the industry works to benefit contractors, who find it advantageous to work with independent adjusters; they ought to be high on a contractor's list of contacts in the insurance industry.

How to Approach Insurance Claims People

It has been said that in order to sell his services, a contractor must first be confident that he is the best at what he does. This advice should be tempered with the knowledge that it is also important to be realistic and honest about one's capabilities. A contractor should thoroughly examine the services he offers and clearly understand his limitations. Promising services one cannot deliver is a sure way to destroy good insurance claims contacts.

The insurance repair contractor should be ready and willing to provide a competent service at a fair price. The contractor must also realize that the adjuster must be allowed to make choices from several contractors.

List and Categorize Contacts
Make a roster of all the insurance claims people to contact. Include not only insurance companies, but also agencies and independent adjusting firms. In addition to the basic information, such as the names, addresses and telephone numbers, the list should include titles and responsibilities, and any other data pertinent to developing a business relationship.

Establish Periodic Contact
Start calling on these individuals periodically, but be careful not to become a nuisance, especially if you have yet to receive the first job or call from someone. Once a contractor knows an agent or claims personnel, the visits take on a more sociable tone. These exchanges are potentially advantageous and positive for both parties because both the contractor's and the client's (insurance agent's) interests are being served. Again, visits should not be

overly frequent. The following example demonstrates what could happen if a contractor inappropriately calls on an agency too often.

A contractor is calling on someone who has yet to give him any work. He visits her every two or three weeks asking for work, yet is never called to produce an estimate. To the objective observer, this contractor is obviously soliciting someone who will never give him a repair job. By continuing to call on that person, he is putting that claims representative in an awkward position. The claims person may have a reason for not calling this contractor. By becoming a nuisance, he forces her to either harshly turn him away or to falsely encourage him with promises that she cannot or will not deliver. Either way, the claims person has a negative impression of the contractor and is not inclined to hire him for any future work.

A visit every two or three months may be appropriate for some prospective clients; others may not like to hear from insurance repair contractors as often. The contractor should exercise careful judgment in making personal visits.

Distributing promotional items is a good way to make contact with claims personnel, rather than coming in just to ask for work. For example, a contractor might distribute a quarterly calendar or message pad imprinted with his advertisement. Every three months, he may stop in to say hello, and distribute these or some other useful items that office personnel need periodically. The object is to gently remind claims personnel that the contractor is out there and can help them if the need arises. If a claims representative gets to know and like a particular contractor, they will call that person for repairs.

Contractors should talk about their services with enthusiasm, and involve their listeners. Ralph Waldo Emerson was right when he said, "Nothing great was ever achieved without enthusiasm." When a contractor demonstrates enthusiasm, claims people will have more confidence that their policyholders will receive prompt, courteous attention and that their insured's damages will be competently repaired.

Contractors seeking insurance repair work should address the claims representative's situation, and describe how their service can be of benefit. The following example illustrates the importance of this technique:

A storekeeper in northern Michigan was annoyed because customers were leaving the doors open in cold weather. He posted signs that read: "For the comfort of others, please close the doors." The doors stayed open and it soon became obvious that the message was not getting across. After some thought, he changed the signs to read, "For your own comfort, please close the doors." The doors were closed.

Make every effort to understand the particular needs of the claims representative, recognizing that those needs vary from company to company. The successful contractor gradually establishes himself as one who provides the service to fill those individual needs.

Use and Develop Good Business Habits
The contractor's business relationship with the community, as well as with the insurance industry, reflects his or her dependability as a solid performer. Participation in service clubs

and associations is a definite advantage. Often this is the best way to develop a personal rapport with local adjusters and agents, since many of them participate in these types of organizations. This is particularly true in smaller communities.

Get to know and be known by people in the insurance community. Be a good listener and take advantage of information obtained from contacts within the industry. Which adjuster is on vacation? Which agency has hired a new claims clerk? Who was promoted? Who has left the company? Who was hired? By becoming personally involved, a contractor may obtain additional business.

Joining Professional Associations

As was previously explained, an important place to meet insurance claims adjusters is through professional associations. Every area of the country has one or two claims associations. Contractors interested in insurance repair work should attempt to become involved in at least one. (A list of claims associations grouped by state is provided in the Appendix.) Be aware that not all adjuster's organizations allow repair contractors to join, but some do and those that do not may still welcome a contractor to attend certain functions and to advertise in their publications.

Advertising

Publications and Directories
Most claims associations have a monthly newsletter or directory. Find out if they sell advertising space, and then consider the cost effectiveness of such advertising. An adjuster will often consult his local claims association publication when looking for a repair firm or expert of some type.

Another excellent tool for advertising is *The Claims Service Guide* published by the Bar List Publishing Co. The *Guide* is distributed to thousands of claims personnel throughout the country and contains extensive listings of contracting firms in cities all over the United States for use by claims people.

Introductory Mail Campaigns
Many repair contractors have successfully penetrated their market with an introductory mail campaign. Begin by obtaining or compiling an adequate mailing list of the claims people in the area. A mailing list may be purchased from claims associations for a fee. Otherwise, compile a list using the market research techniques discussed in previous sections of this chapter.

When designing the material for a mailing, remember that it is absolutely vital to grab the adjuster's attention. It is not enough for a repairman to send a letter stating what he can do, because *every* repairman says the same thing. Include a "freebie" – an item that would be useful to an adjuster, such as a roofer's pitch card or a rafter length table. Getting something that is clearly useful will greatly improve the chances that the adjuster will read the accompanying letter and retain the contractor's business card for future work.

Promotional Give-aways
Promotional "give-aways" are of no value unless they have a function and are truly useful to an adjuster. Only if he has use

for an item will it stay on his desk, in his sight. By all means, make sure the company name, telephone number, and (if possible) logo are prominently displayed on the item.

The following is a list of items that have particular appeal to adjusters.

- Flashlights
- Flashlight Pens
- Roof Pitch Cards
- Rafter Length Tables
- Rulers
- Construction Templates
- Memo Pads
- Pens and Pencils
- Coffee Mugs

Numerous companies supply promotional items and, for a fee, imprint a company name or an advertising message. Two such firms are Sales Guides, Inc. of Mequon, Wisconsin, and Atlas Pen & Pencil Corporation of Hollywood, Florida.

No matter how personable a contractor is or how much money he or she spends on promotional items and advertising, the service he/she provides is the most important credential to attract continuing business. Pricing and competence are fundamental. Furthermore, a contractor who maintains high quality work and who delivers what he/she promises establishes a profitable ongoing relationship with insurance adjusters.

Chapter 4

HOW TO WRITE ESTIMATES FOR ADJUSTERS

In estimating damage repair costs, insurance companies require that adjusters achieve the highest degree of accuracy possible. Taking correct measurements, properly identifying materials, understanding labor requirements, and making accurate computations are crucial to this endeavor. To succeed, the adjuster needs a systematic means or manner of procedure – a method. This method should allow the estimate to be critiqued and compared "room by room" with a competing bid.

The Estimating Process

The insurance repair contractor should develop a workable estimating method, and should use this method uniformly for all estimates. Following an established process streamlines a contractor's activities, minimizes loss of time, and eliminates mistakes in the final bid. The step-by-step process includes preparing a rough diagram of the damaged area, taking measurements, calculating areas of each room, and presenting a final, typed takeoff.

Job Orientation
The contractor prepares a rough diagram of the structure, carefully recording the measurements of each room. Although attention to the specifics of the damages is not necessary at this point, the contractor should become somewhat familiar with the extent of damage to various structural items. The contractor should also learn as much as possible about the site and the nature of required repairs. He should review "as-built" drawings (if available), and should take pictures and notes.

Quantity Takeoff
After the rough diagram and measurements are complete, the contractor records a room-by-room quantity takeoff. The contractor should record in detail every repair required as a result of the insured loss for each room: how many square feet of drywall, lineal feet of studs, yards of carpet, etc. This quantity takeoff (also known as a *general scope*) must be thorough and accurate, because it is the document used to prepare the final bid. A pre-printed form, as shown in Figure 4.1 (from *Means Forms for Building Construction Professionals*) can be used to organize the information collected during the measurement process.

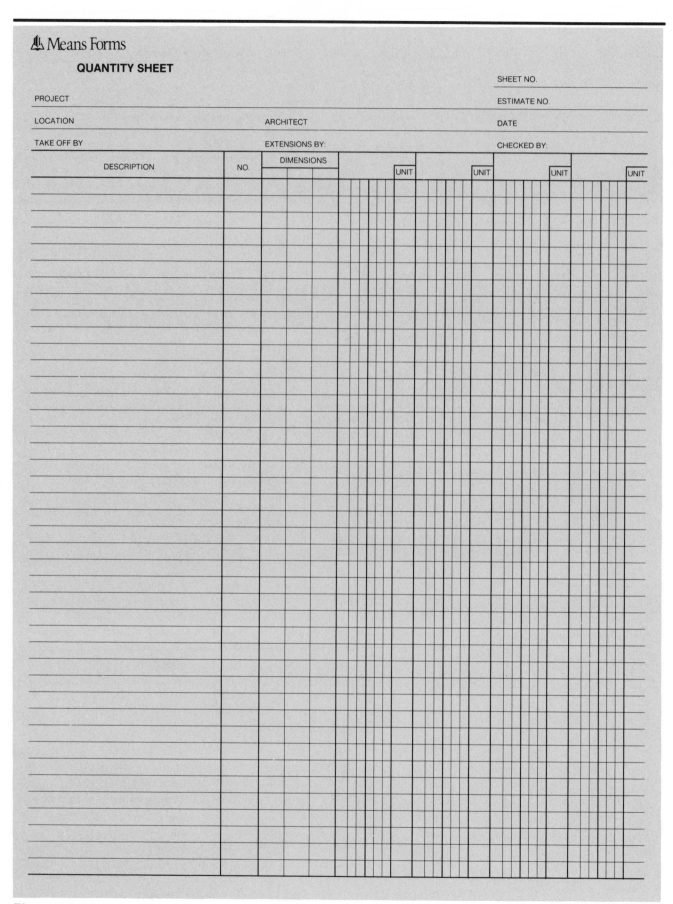

Figure 4.1

The next step is to estimate quantities of materials needed to perform the work (for example, the lineal feet of base moldings, or the square feet of sub-flooring).

Pricing the Estimate

From the information collected during the takeoff, prices can be determined for the repairs. Published cost guides such as *Means Repair and Remodeling Cost Data*, Means *Building Construction Cost Data*, or the contractor's own historical cost records can be used. The final estimate should clearly state the quantities, unit prices, subtotals, and an added percentage for overhead and profit.

Finally, the estimate is clearly printed or typed in an orderly, professional layout that will be easy for the adjuster to understand and utilize. A standard form, such as Means Cost Analysis form (shown in Figure 4.2) provides a consistent format for organizing this data.

Upon completion, the contractor should immediately deliver the estimate to the insurance adjuster. Some contractors may provide the policyholder with a copy of the estimate as well. However, always consult the adjuster before giving a copy of the estimate to the policyholder.

Practicing this step-by-step method is a good way for contractors to improve their estimating skills. It is also a systematic way to analyze questionable areas (such as possible unseen damage to wall structures).

Taking Measurements

The insurance repair contractor should understand the basic difference between estimating for new construction and estimating damage repairs. For new construction, plans and specifications listing complete measurements are issued to several competing contractors, who use this data to prepare a takeoff and bid.

The situation is quite different for contractors estimating damaged structures. For damage repairs, the contractor must create specifications at the site by recording accurate measurements throughout the structure.

In some situations, it is more difficult to compute quantities or areas. For example, what is the area of the gable sides of a building? What is the area of a hip roof? To determine this visually, segment the roof into four triangles, compute the area of each triangle and then multiply by four to get the total area. Some commonly-used formulas appear in the Appendix.

An insurance repair contractor should always carry a tape measure to every prospective job site. A 25' steel tape is recommended for taking interior measurements and a 100' steel or flexible tape for longer exterior measurements. A measuring wheel or other similar device may also be used.

When recording the dimensions of a room, some adjusters and contractors prefer to round measurements up or down to the nearest whole or half foot. With this method, a room with true dimensions of 10'-2" x 15'-4" x 8' may be recorded in the estimate as 10' x 15'-6" x 8'. It is generally better to record exact measurements so that the final estimate is more precise and,

Means Forms

COST ANALYSIS

SHEET NO. _____

PROJECT _____ ESTIMATE NO. _____

ARCHITECT _____ DATE _____

TAKE OFF BY: _____ QUANTITIES BY: _____ PRICES BY: _____ EXTENSIONS BY: _____ CHECKED BY: _____

DESCRIPTION	SOURCE/DIMENSIONS			QUANTITY	UNIT	MATERIAL		LABOR		EQ./TOTAL	
						UNIT COST	TOTAL	UNIT COST	TOTAL	UNIT COST	TOTAL

Figure 4.2

52

therefore, less likely to be criticized or challenged. (Guidelines for converting fractional footage to decimal numbers are provided later in this chapter.)

Note all units of measure used to calculate areas and quantities recorded. Be sure to write down the length, width, and height for each room to be painted or plastered. Give the exact measurements of roof, wall, or floor areas to be replaced. Specify the dimensions of building items that are sold in specified lengths, widths, or thicknesses (such as doors and windows); for example, *a standard door of 6'-8" x 3'*.

Figure 4.3 is an example of a standard form used to record the measurements of an insurance repair job. In this example, only a small area of the home was affected by a fire, which started in the kitchen when some overheated oil caught fire, which then spread to a nearby cabinet. Smoke and soot affected the other rooms listed, but the homeowner was able to put out the fire before it burned any areas outside of the kitchen.

Floor Plan

A floor plan should be prepared – at least for the portion of the structure that was damaged. The plan may be drawn on standard graph paper and should be attached to the formal estimate. A sample partial diagram is shown in Figure 4.4.

Perimeter Diagram

Occasionally, a contractor prepares a diagram of the structure's perimeter for the benefit of the adjuster. Insurance companies need this information in order to determine the values of buildings they insure. Even if the adjuster has already prepared a valuation himself, he may still appreciate the effort and view such a diagram as a "bonus" provided by the contractor. A sample perimeter (or *peripheral*) diagram is shown in Figure 4.5.

Converting Fractional Footage to Decimals

When measurements are in whole feet, the mathematics are simple. Suppose an area to be painted measures 80' x 16' and the unit price is .32 cents per square foot for painting. In order to find the total square footage of the area, multiply 80' x 16', which equals 1,280 square feet. This figure is multiplied by .32 to arrive at the total price, $409.60. This calculation is shown in Figure 4.6.

When measurements involve fractions, the mathematics are more complicated. Suppose that the area to be painted is 16'-6" x 79'-3". How do you determine the square footage to multiply by the 32 cents? Rounding off to the nearest square footage would be either 16' x 79' or 17' x 79'. If these figures were used, the overall estimate would be far from accurate.

One solution is to write out the math problem in fractions, as shown in Figure 4.7. This method can be complicated and time-consuming, especially considering the frequency with which rooms or areas do not measure in whole feet.

A much easier method, which is readily accepted by adjusters, is to convert fractions to decimals. Decimals are easy to work with, and require only an inexpensive calculator. Measure as usual and record the measurement in feet and inches, for example, 8'-6" x 9'-3". The chart in Figure 4.8 can then be used to convert the inches into decimal equivalents. The fractional footage is

QUANTITY SHEET

SHEET NO. _1_

PROJECT _Fire Loss – 457 Beechwood Ave._

ESTIMATE NO.

LOCATION _Miami, Florida_   ARCHITECT _N/A_

DATE

TAKE OFF BY _J. Contractor_   EXTENSIONS BY:

CHECKED BY:

DESCRIPTION	NO.	DIMENSIONS			UNIT		UNIT		UNIT		UNIT
Kitchen (8' height)		9'	×	7'4"							
– base cabinets 36" ht.		48"	×	24"							
– single hung aluminum window	1	42"	×	42"							
Pantry (8' height)		3'	×	2'							
– hollow core door	1	2'6"	×	72"							
Living Room (8' height)		14'	×	18'							
– wall covering / one wall		14'	×	8'							
Dining Room (8' height)		14'6"	×	10'							
Hall (8' height)		3'6"	×	16'							
Closet (8' height)		2'	×	2'							

Figure 4.3

converted to decimals as follows: 8.5 x 9.25 feet, and multiplied to get the total in square feet (8.5 x 9.25 feet = 79 S.F.). Then multiply the area by the price per square foot (79 S.F. x .32 per square foot = $ 25.28), as shown in Figure 4.9.

Square Foot Tables

The Square Foot Tables shown in Figure 4.10 were designed to allow easy calculation of wall and floor areas in normal-sized rooms. For example, to quickly determine the total square footage of the walls and ceiling in a room measuring 18' x 16' with a height of 8', refer to the heading "For Rooms with Ceilings 8 Feet High" in Figure 4.10. Cross-referencing the horizontal and diagonal numbers, the total area of the four walls plus the ceiling is 832 square feet.

General Scope or Quantity Takeoff

Once the job orientation and the exact quantities and types of materials to be used have been itemized, the next step is the quantity takeoff. Some contractors prefer to dictate these

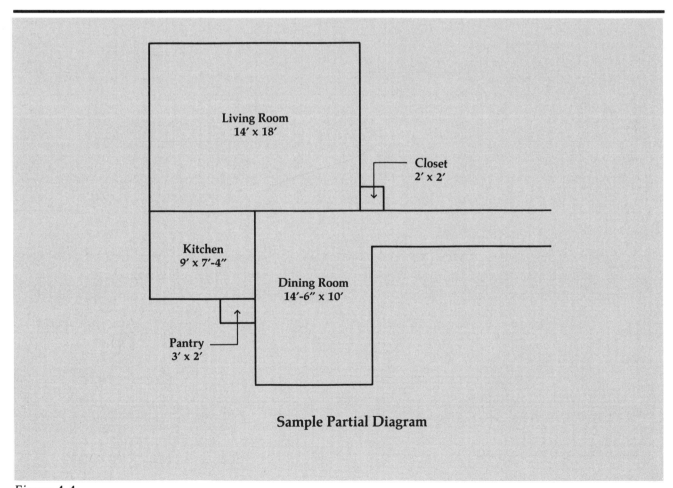

Sample Partial Diagram

Figure 4.4

specifications into a cassette recorder while they are on the site, and have them typed at a later time. Most simply take notes.

Note-taking should begin in the room where the loss originated. Do not leave this room or area until every last detail necessary to restore the property to its original condition has been recorded. Never record part of the loss in one area, move to another area, and then return to the first. This approach results in an unprofessional and disorganized final estimate, and omissions are more likely to occur. Before moving to the next area, room, or floor, be sure the takeoff is complete.

Many contractors use a prepared checklist to record damaged items and the repairs needed. A typical checklist for insurance repair projects is shown in Figure 4.11. This checklist generally follows the MASTERFORMAT system of classification and numbering, developed by the Construction Specifications Institute (CSI). Since the Means unit cost guides and many product catalogues also follow this format, the use of this checklist will facilitate the pricing of the quantity takeoff. Some deviation from the MASTERFORMAT may be necessary, however, to conform

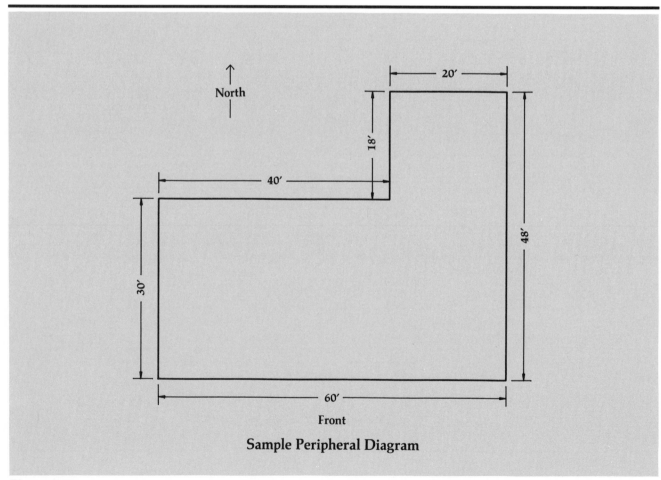

Sample Peripheral Diagram

Figure 4.5

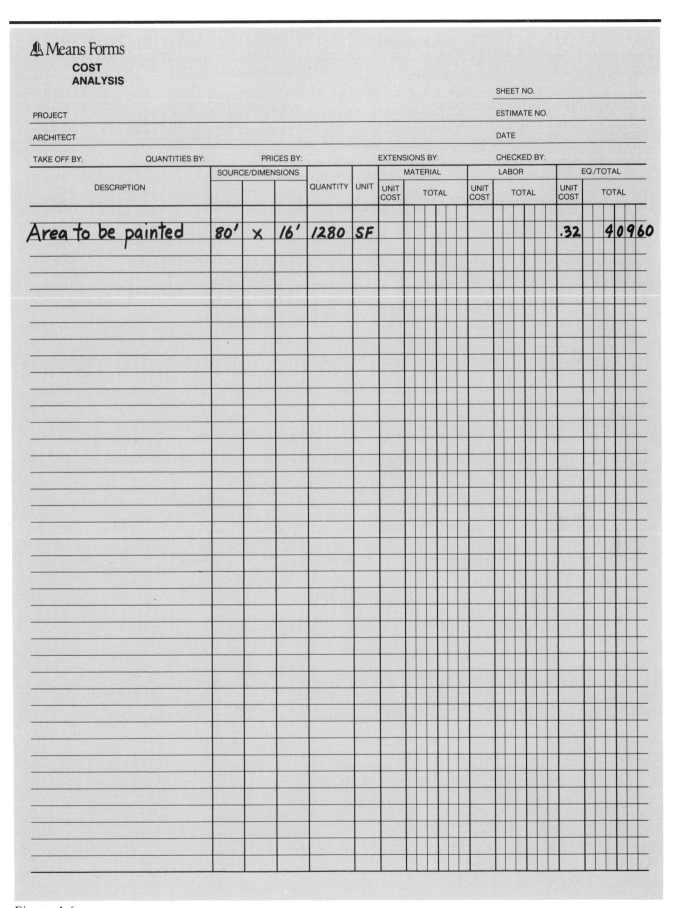

Figure 4.6

57

with insurance industry custom. Be alert to any special items specific to that job that are not on the checklist. The checklist is the "foundation" of an orderly estimate. Each category on the checklist is discussed in the following sections.

Tear-Out and Debris Removal

Before new materials can be installed in a structure to be repaired, the contractor must remove and dispose of all of the damaged materials. Labor production quantities from published cost guides, such as Means *Building Construction Cost Data* or *Means Repair & Remodeling Cost Data* can be used to approximate the amount of tear-out and debris removal. These reference guides include information on labor requirements for particular tear-out operations. For example, to determine the labor involved in removing 4,000 S.F. of roofing, published cost guides provide the number of square feet a crew could remove in one hour, and the daily output level for that crew. Although a contractor's data is the ideal source of such productivity information, annually-updated guides such as these can be very useful, especially if a contractor's pricing experience is limited.

Taking off and estimating tear-out and debris removal involves the costs of labor plus any specialized equipment (rented or purchased). Many estimators make the mistake of guessing at the overall cost of tear-out and debris removal rather than analyzing each single operation. These overall guesses may be too high, in which case the adjuster may assume that the contractor is using the tear-out item to "bury" unexpected costs or excess profits.

To estimate labor costs, some contractors picture each individual operation, and simply estimate the number of man-hours required to perform the task based on experience. Another alternative is to use a published guide for demolition output.

$$16 \frac{6}{12} \times 79 \frac{3}{12} = \frac{198}{12} \times \frac{951}{12}$$

$$\frac{198}{12} \times \frac{951}{12} = \frac{188,298}{144}$$

$$\frac{188,298}{144} = 1,307.63 \text{ SF}$$

Figure 4.7

Figure 4.12 is an excerpt from Means *Repair and Remodeling Cost Data*, 1989. It shows the daily output for a crew engaged in removing concrete plank roof decking.

Using the measurements from the handwritten takeoff (shown earlier in Figure 4.3), and comparing output figures from the Means cost guide, the tear-out and debris removal is computed for the kitchen in Figure 4.13.

Copies of quantity takeoff sheets are generally kept in the contractor's office so that, if asked, he can produce a document from the file showing how the quantities were determined.

Undamaged Item Removal

Many times, undamaged materials must be removed in order to gain access to the damaged materials. The costs involved can be surprising. For example: finished flooring must be removed in order to replace damaged floor joists, or an undamaged sink must be removed from a burned cabinet and reinstalled in a new cabinet. Machinery, appliances, or other property may have to be moved in order to make repairs to a floor. The repair contractor should carefully inspect even undamaged areas and items for costs associated with removing and, possibly, replacing them.

Decimal Parts of a Foot		
Major Divisions	**Secondary Division**	**For added 1/8" Increments**
0 = 0.00		1/8" = .01
		1/4" = .02
	1" = 0.08	3/8" = .03
		1/2" = .04
	2" = 0.17	5/8" = .05
		3/4" = .06
3" = 0.25		7/8" = .07
6" = 0.50	4" = 0.33	
	5" = 0.42	
9" = 0.75	7" = 0.58	
	8" = 0.67	
12" = 1.00	10" = 0.83	
	11" = 0.92	

Thus, 3' – 7 3/4" becomes.

3'	=	3.00
7	=	.58
3/4"	=	.06
		3.64 feet

Figure 4.8

Means Forms

COST ANALYSIS

			SHEET NO.
PROJECT			ESTIMATE NO.
ARCHITECT			DATE

TAKE OFF BY:	QUANTITIES BY:	PRICES BY:	EXTENSIONS BY:	CHECKED BY:

DESCRIPTION	SOURCE/DIMENSIONS			QUANTITY	UNIT	MATERIAL		LABOR		EQ./TOTAL	
						UNIT COST	TOTAL	UNIT COST	TOTAL	UNIT COST	TOTAL
Sample measurement	8.5	x	9.25	79	SF					.32	25 28

Figure 4.9

Rooms with 7' Ceilings—Total Area – 4 Walls and Ceiling (in square feet)

	3'	4'	5'	6'	7'	8'	9'	10'	11'	12'	13'	14'	15'	16'	17'	18'	19'	20'	21'	22'
3'	93	110	127	144	161	178	195	212	229	246	263	280	297	314	331	348	365	382	399	416
4'	110	128	146	164	182	200	218	236	254	272	290	308	326	344	362	380	398	416	434	452
5'	127	146	165	184	203	222	241	260	279	298	317	336	355	374	393	412	431	450	469	488
6'	144	164	184	204	224	244	264	284	304	324	344	364	384	404	424	444	464	484	504	524
7'	161	182	203	224	245	266	287	308	329	350	371	392	413	434	455	476	497	518	539	560
8'	178	200	222	244	266	288	310	332	354	376	398	420	442	464	486	508	530	552	574	596
9'	195	218	241	264	287	310	333	356	379	402	425	448	471	494	517	540	563	586	609	632
10'	212	236	260	284	308	332	356	380	404	428	452	476	500	524	548	572	596	620	644	668
11'	229	254	279	304	329	354	379	404	429	454	479	504	529	554	579	604	629	654	679	704
12'	246	272	298	324	350	376	402	428	454	480	506	532	558	584	610	636	662	688	714	740
13'	263	290	317	344	371	398	425	452	479	506	533	560	587	614	641	668	695	722	749	776
14'	280	308	336	364	392	420	448	476	504	532	560	588	616	644	672	700	728	756	784	812
15'	297	326	355	384	413	442	471	500	529	558	587	616	645	674	703	732	761	790	819	848
16'	314	344	374	404	434	464	494	524	554	584	614	644	674	704	734	764	794	824	854	884
17'	31	362	393	424	455	486	517	548	579	610	641	672	703	734	765	796	827	858	889	920
18'	348	380	412	444	476	508	540	572	604	636	668	700	732	764	796	828	860	892	924	956
19'	365	398	431	464	497	530	563	596	629	662	695	728	761	794	827	860	893	926	959	992
20'	382	416	450	484	518	552	586	620	654	688	722	756	790	824	858	892	926	960	994	1028
21'	399	434	469	504	539	574	609	644	679	714	749	784	819	854	889	924	959	994	1029	1064
22'	416	452	488	524	560	596	632	668	704	740	776	812	848	884	920	956	992	1028	1064	1100
23'	433	470	507	544	581	618	655	692	729	766	803	840	877	914	951	988	1025	1062	1099	1136
24'	450	488	526	564	602	640	678	716	754	792	830	868	906	944	982	1020	1058	1096	1134	1172

Rooms with 7'6" Ceilings

	3'	4'	5'	6'	7'	8'	9'	10'	11'	12'	13'	14'	15'	16'	17'	18'	19'	20'	21'	22'
3'	99	117	135	153	171	189	207	225	243	261	279	297	315	333	351	369	387	405	423	441
4'	117	136	155	174	193	212	231	250	269	288	307	326	345	364	383	402	421	440	459	478
5'	135	155	175	195	215	235	255	275	295	315	335	355	375	395	415	435	455	475	495	515
6'	153	174	195	216	237	258	279	300	321	342	363	384	405	426	447	468	489	510	531	552
7'	171	193	215	237	259	281	303	325	347	369	391	413	435	457	479	501	523	545	567	589
8'	189	212	235	258	281	304	327	350	373	396	419	442	465	488	511	534	557	580	603	626
9'	207	231	255	279	303	327	351	375	399	423	447	471	495	519	543	567	591	615	639	663
10'	225	250	275	300	325	350	375	400	425	450	475	500	525	550	575	600	625	650	675	700
11'	243	269	295	321	347	373	399	425	451	477	503	529	555	581	607	633	659	685	711	737
12'	261	288	315	342	369	396	423	450	477	504	531	558	585	612	639	666	693	720	747	774
13'	279	307	335	363	391	419	447	475	503	531	559	587	615	643	671	699	727	755	783	811
14'	297	326	355	384	413	442	471	500	529	558	587	616	645	674	703	732	761	790	819	848
15'	315	345	375	405	435	465	495	525	555	585	615	645	675	705	735	765	795	825	855	885
16'	333	364	395	426	457	488	519	550	581	612	643	674	705	736	767	798	829	860	891	922
17'	351	383	415	447	479	511	543	575	607	639	671	703	735	767	799	831	863	895	927	959
18'	369	402	435	468	501	534	567	600	633	666	699	732	765	798	831	864	897	930	963	996
19'	387	421	455	489	523	557	591	625	659	693	727	761	795	829	863	897	931	965	999	1033
20'	405	440	475	510	545	580	615	650	685	720	755	790	825	860	895	930	965	1000	1035	1070
21'	423	459	495	531	567	603	639	675	711	747	783	819	855	891	927	963	999	1035	1071	1107
22'	441	478	515	552	589	626	663	700	737	774	811	848	885	922	959	996	1033	1070	1107	1144
23'	459	497	535	573	611	649	687	725	763	801	839	877	915	953	991	1029	1067	1105	1143	1181
24'	477	516	555	594	633	672	711	750	789	828	867	906	945	984	1023	1062	1101	1140	1179	1218

Rooms with 8' Ceilings

	3'	4'	5'	6'	7'	8'	9'	10'	11'	12'	13'	14'	15'	16'	17'	18'	19'	20'	21'	22'
3'	105	124	143	162	181	200	219	238	257	276	295	314	333	352	371	390	409	428	447	466
4'	124	144	164	184	204	224	244	264	284	304	324	344	364	384	404	424	444	464	484	504
5'	143	164	185	206	227	248	269	290	311	332	353	374	395	416	437	458	479	500	521	542
6'	162	184	206	228	250	272	294	316	338	360	382	404	426	448	470	492	514	536	558	580
7'	181	204	227	250	273	296	319	342	365	388	411	434	457	480	503	526	549	572	595	618
8'	200	224	248	272	296	320	344	368	392	416	440	464	488	512	536	560	584	608	632	656
9'	219	244	269	294	319	344	369	394	419	444	469	494	519	544	569	594	619	644	669	694
10'	238	264	290	316	342	368	394	420	446	472	498	524	550	576	602	628	654	680	706	732
11'	257	284	311	338	365	392	419	446	473	500	527	554	581	608	635	662	689	716	743	770
12'	276	304	332	360	388	416	444	472	500	528	556	584	612	640	668	696	724	752	780	808
13'	295	324	353	382	411	440	469	498	527	556	585	614	643	672	701	730	759	788	817	846
14'	314	344	374	404	434	464	494	524	554	584	614	644	674	704	734	764	794	824	854	884
15'	333	364	395	426	457	488	519	550	581	612	643	674	705	736	767	798	829	860	891	922
16'	352	384	416	448	480	512	544	576	608	640	672	704	736	768	800	832	864	896	928	960
17'	371	404	437	470	503	536	569	602	635	668	701	734	767	800	833	866	899	932	965	998
18'	390	424	458	492	526	560	594	628	662	696	730	764	798	832	866	900	934	968	1002	1036
19'	409	444	479	514	549	584	619	654	689	724	759	794	829	864	899	934	969	1004	1039	1074
20'	428	464	500	536	572	608	644	680	716	752	788	824	860	896	932	968	1004	1040	1076	1112
21'	447	484	521	558	595	632	669	706	743	780	817	854	891	928	965	1002	1039	1076	1113	1150
22'	466	504	542	580	618	656	694	732	770	808	846	884	922	960	998	1036	1074	1112	1150	1188
23'	485	524	563	602	641	680	719	758	797	836	875	914	953	992	1031	1070	1109	1148	1187	1226
24'	504	544	584	624	664	704	744	784	824	864	904	944	984	1024	1064	1104	1144	1184	1224	1264

Figure 4.10

Rooms with 8'6" Ceilings—Total Area - 4 Walls and Ceiling (in square feet)

	3'	4'	5'	6'	7'	8'	9'	10'	11'	12'	13'	14'	15'	16'	17'	18'	19'	20'	21'	22'
3'	111	131	151	171	191	211	231	251	271	291	311	331	351	371	391	411	431	451	471	491
4'	131	152	173	194	215	236	257	278	299	320	341	362	383	404	425	446	467	488	509	530
5'	151	173	195	217	239	261	283	305	327	349	371	393	415	437	459	481	503	525	547	569
6'	171	194	217	240	263	286	309	332	355	378	401	424	447	470	493	516	539	562	585	608
7'	191	215	239	263	287	311	335	359	383	407	431	455	479	503	527	551	575	599	623	647
8'	211	236	261	286	311	336	361	386	411	436	461	486	511	536	561	586	611	636	661	686
9'	231	257	283	309	335	361	387	413	439	465	491	517	543	569	595	621	647	673	699	725
10'	251	278	305	332	359	386	413	440	467	494	521	548	575	602	629	656	683	710	737	764
11'	271	299	327	355	383	411	439	467	495	523	551	579	607	635	663	691	719	747	775	803
12'	291	320	349	378	407	436	465	494	523	552	581	610	639	668	697	726	755	784	813	842
13'	311	341	371	401	431	461	491	521	551	581	611	641	671	701	731	761	791	821	851	881
14'	331	362	393	424	455	486	517	548	579	610	641	672	703	734	765	796	827	858	889	920
15'	351	383	415	447	479	511	543	575	607	639	671	703	735	767	799	831	863	895	927	959
16'	371	404	437	470	503	536	569	602	635	668	701	734	767	800	833	866	899	932	965	998
17'	391	425	459	493	527	561	595	629	663	697	731	765	799	833	867	901	935	969	1003	1037
18'	411	446	481	516	551	586	621	656	691	726	761	796	831	866	901	936	971	1006	1041	1076
19'	431	467	503	539	575	611	647	683	719	755	791	827	863	899	935	971	1007	1043	1079	1115
20'	451	488	525	562	599	636	673	710	747	784	821	858	895	932	969	1006	1043	1080	1117	1154
21'	471	509	547	585	623	661	699	737	775	813	851	889	927	965	1003	1041	1079	1117	1155	1193
22'	491	530	569	608	647	686	725	764	803	842	881	920	959	998	1037	1076	1115	1154	1193	1232
23'	511	551	591	631	671	711	751	791	831	871	911	951	991	1031	1071	1111	1151	1191	1231	1271
24'	531	572	613	654	695	736	777	818	859	900	941	982	1023	1064	1105	1146	1187	1228	1269	1310

Rooms with 9' Ceilings

	3'	4'	5'	6'	7'	8'	9'	10'	11'	12'	13'	14'	15'	16'	17'	18'	19'	20'	21'	22'
3'	117	138	159	180	201	222	243	264	285	306	327	348	369	390	411	432	453	474	495	516
4'	138	160	182	204	226	248	270	292	314	336	358	380	402	424	446	468	490	512	534	556
5'	159	182	205	228	251	274	297	320	343	366	389	412	435	458	481	504	527	550	573	596
6'	180	204	228	252	276	300	324	348	372	396	420	444	468	492	516	540	564	588	612	636
7'	201	226	251	276	301	326	351	376	401	426	451	476	501	526	551	576	601	626	651	676
8'	222	248	274	300	326	352	378	404	430	456	482	508	534	560	586	612	638	664	690	716
9'	243	270	297	324	351	378	405	432	459	486	513	540	567	594	621	648	675	702	729	756
10'	264	292	320	348	376	404	432	460	488	516	544	572	600	628	656	684	712	740	768	796
11'	285	314	343	372	401	430	459	488	517	546	575	604	633	662	691	720	749	778	807	836
12'	306	336	366	396	426	456	486	516	546	576	606	636	666	696	726	756	786	816	846	876
13'	327	358	389	420	451	482	513	544	575	606	637	668	699	730	761	792	823	854	885	916
14'	348	380	412	444	476	508	540	572	604	636	668	700	732	764	796	828	860	892	924	956
15'	369	402	435	468	501	534	567	600	633	666	699	732	765	798	831	864	897	930	963	996
16'	390	424	458	492	526	560	594	628	662	696	730	764	798	832	866	900	934	968	1002	1036
17'	411	446	481	516	551	586	621	656	691	726	761	796	831	866	901	936	971	1006	1041	1076
18'	432	468	504	540	576	612	648	684	720	756	792	828	864	900	936	972	1008	1044	1080	1116
19'	453	490	527	564	601	638	675	712	749	786	823	860	897	934	971	1008	1045	1082	1119	1156
20'	474	512	550	588	626	664	702	740	778	816	854	892	930	968	1006	1044	1082	1120	1158	1196
21'	495	534	573	612	651	690	729	768	807	846	885	924	963	1002	1041	1080	1119	1158	1197	1236
22'	516	556	596	636	676	716	756	796	836	876	916	956	996	1036	1076	1116	1156	1196	1236	1276
23'	537	578	619	660	701	742	783	824	865	906	947	988	1029	1070	1111	1152	1193	1234	1275	1316
24'	558	600	642	684	726	768	810	852	894	936	978	1020	1062	1104	1146	1188	1230	1272	1314	1356

Rooms with 9'6" Ceilings

	3'	4'	5'	6'	7'	8'	9'	10'	11'	12'	13'	14'	15'	16'	17'	18'	19'	20'	21'	22'
3'	123	145	167	189	211	233	255	277	299	321	343	365	387	409	431	453	475	497	519	541
4'	145	168	191	214	237	260	283	306	329	352	375	398	421	444	467	490	513	536	559	582
5'	167	191	215	239	263	287	311	335	359	383	407	431	455	479	503	527	551	575	599	623
6'	189	214	239	264	289	314	339	364	389	414	439	464	489	514	539	564	589	614	639	664
7'	211	237	263	289	315	341	367	393	419	445	471	497	523	549	575	601	627	653	679	705
8'	233	260	287	314	341	368	395	422	449	476	503	530	557	584	611	638	665	692	719	746
9'	255	283	311	339	367	395	423	451	479	507	535	563	591	619	647	675	703	731	759	787
10'	277	306	335	364	393	422	451	480	509	538	567	596	625	654	683	712	741	770	799	828
11'	299	329	359	389	419	449	479	509	539	569	599	629	659	689	719	749	779	809	839	869
12'	321	352	383	414	445	476	507	538	569	600	631	662	693	724	755	786	817	848	879	910
13'	343	375	407	439	471	503	535	567	599	631	663	695	727	759	791	823	855	887	919	951
14'	365	398	431	464	497	530	563	596	629	662	695	728	761	794	827	860	893	926	959	992
15'	387	421	455	489	523	557	591	625	659	693	727	761	795	829	863	897	931	965	999	1033
16'	409	444	479	514	549	584	619	654	689	724	759	794	829	864	899	934	969	1004	1039	1074
17'	431	467	503	539	575	611	647	683	719	755	791	827	863	899	935	971	1007	1043	1079	1115
18'	453	490	527	564	601	638	675	712	749	786	823	860	897	934	971	1008	1045	1082	1119	1156
19'	475	513	551	589	627	665	703	741	779	817	855	893	931	969	1007	1045	1083	1121	1159	1197
20'	497	536	575	614	653	692	731	770	809	848	887	926	965	1004	1043	1082	1121	1160	1199	1238
21'	519	559	599	639	679	719	759	799	839	879	919	959	999	1039	1079	1119	1159	1199	1239	1279
22'	541	582	623	664	705	746	787	828	869	910	951	992	1033	1074	1115	1156	1197	1238	1279	1320
23'	563	605	647	689	731	773	815	857	899	941	983	1025	1067	1109	1151	1193	1235	1277	1319	1361
24'	585	628	671	714	757	800	843	886	929	972	1015	1058	1101	1144	1187	1230	1273	1316	1359	1402

Figure 4.10 (cont.)

Rooms with 10′ Ceilings—Total Area – 4 Walls and Ceiling (in square feet)

	3′	4′	5′	6′	7′	8′	9′	10′	11′	12′	13′	14′	15′	16′	17′	18′	19′	20′	21′	22′
3′	129	152	175	198	221	244	267	290	313	336	359	382	405	428	451	474	497	520	543	566
4′	152	176	200	224	248	272	296	320	344	368	392	416	440	464	488	512	536	560	584	608
5′	175	200	225	250	275	300	325	350	375	400	425	450	475	500	525	550	575	600	625	650
6′	198	224	250	276	302	328	354	380	406	432	458	484	510	536	562	588	614	640	666	692
7′	221	248	275	302	329	356	383	410	437	464	491	518	545	572	599	626	653	680	707	734
8′	244	272	300	328	356	384	412	440	468	496	524	552	580	608	636	664	692	720	748	776
9′	267	296	325	354	383	412	441	470	499	528	557	586	615	644	673	702	731	760	789	818
10′	290	320	350	380	410	440	470	500	530	560	590	620	650	680	710	740	770	800	830	860
11′	313	344	375	406	437	468	499	530	561	592	623	654	685	716	747	778	809	840	871	902
12′	336	368	400	432	464	496	528	560	592	624	656	688	720	752	784	816	848	880	912	944
13′	359	392	425	458	491	524	557	590	623	656	689	722	755	788	821	854	887	920	953	986
14′	382	416	450	484	518	552	586	620	654	688	722	756	790	824	858	892	926	960	994	1028
15′	405	440	475	510	545	580	615	650	685	720	755	790	825	860	895	930	965	1000	1035	1070
16′	428	464	500	536	572	608	644	680	716	752	788	824	860	896	932	968	1004	1040	1076	1112
17′	451	488	525	562	599	636	673	710	747	784	821	858	895	932	969	1006	1043	1080	1117	1154
18′	474	512	550	588	626	664	702	740	778	816	854	892	930	968	1006	1044	1082	1120	1158	1196
19′	497	536	575	614	653	692	731	770	809	848	887	926	965	1004	1043	1082	1121	1160	1199	1238
20′	520	560	600	640	680	720	760	800	840	880	920	960	1000	1040	1080	1120	1160	1200	1240	1280
21′	543	584	625	666	707	748	789	830	871	912	953	994	1035	1076	1117	1158	1199	1240	1281	1322
22′	566	608	650	692	734	776	818	860	902	944	986	1028	1070	1112	1154	1196	1238	1280	1322	1364
23′	589	632	675	718	761	804	847	890	933	976	1019	1062	1105	1148	1191	1234	1277	1320	1363	1406
24′	612	656	700	744	788	832	876	920	964	1008	1052	1096	1140	1184	1228	1272	1316	1360	1404	1448

Rooms with 10′6″ Ceilings

	3′	4′	5′	6′	7′	8′	9′	10′	11′	12′	13′	14′	15′	16′	17′	18′	19′	20′	21′	22′
3′	135	159	183	207	231	255	279	303	327	351	375	399	423	447	471	495	519	543	567	591
4′	159	184	209	234	259	284	309	334	359	384	409	434	459	484	509	534	559	584	609	634
5′	183	209	235	261	287	313	339	365	391	417	443	469	495	521	547	573	599	625	651	677
6′	207	234	261	288	315	342	369	396	423	450	477	504	531	558	585	612	639	666	693	720
7′	231	259	287	315	343	371	399	427	455	483	511	539	567	595	623	651	679	707	735	763
8′	255	284	313	342	371	400	429	458	487	516	545	574	603	632	661	690	719	748	777	806
9′	279	309	339	369	399	429	459	489	519	549	579	609	639	669	699	729	759	789	819	849
10′	303	334	365	396	427	458	489	520	551	582	613	644	675	706	737	768	799	830	861	892
11′	327	359	391	423	455	487	519	551	583	615	647	679	711	743	775	807	839	871	903	935
12′	351	384	417	450	483	516	549	582	615	648	681	714	747	780	813	846	879	912	945	978
13′	375	409	443	477	511	545	579	613	647	681	715	749	783	817	851	885	919	953	987	1021
14′	399	434	469	504	539	574	609	644	679	714	749	784	819	854	889	924	959	994	1029	1064
15′	423	459	495	531	567	603	639	675	711	747	783	819	855	891	927	963	999	1035	1071	1107
16′	447	484	521	558	595	632	669	706	743	780	817	854	891	928	965	1002	1039	1076	1113	1150
17′	471	509	547	585	623	661	699	737	775	813	851	889	927	965	1003	1041	1079	1117	1155	1193
18′	495	534	573	612	651	690	729	768	807	846	885	924	963	1002	1041	1080	1119	1158	1197	1236
19′	519	559	599	639	679	719	759	799	839	879	919	959	999	1039	1079	1119	1159	1199	1239	1279
20′	543	584	625	666	707	748	789	830	871	912	953	994	1035	1076	1117	1158	1199	1240	1281	1322
21′	567	609	651	693	735	777	819	861	903	945	987	1029	1071	1113	1155	1197	1239	1281	1323	1365
22′	591	634	677	720	763	806	849	892	935	978	1021	1064	1107	1150	1193	1236	1279	1322	1365	1408
23′	615	659	703	747	791	835	879	923	967	1011	1055	1099	1143	1187	1231	1275	1319	1363	1407	1451
24′	639	684	729	774	819	864	909	954	999	1044	1089	1134	1179	1224	1269	1314	1359	1404	1449	1494

Rooms with 11′ Ceilings

	3′	4′	5′	6′	7′	8′	9′	10′	11′	12′	13′	14′	15′	16′	17′	18′	19′	20′	21′	22′
3′	141	166	191	216	241	266	291	316	341	366	391	416	441	466	491	516	541	566	591	616
4′	166	192	218	244	270	296	322	348	374	400	426	452	478	504	530	556	582	608	634	660
5′	191	218	245	272	299	326	353	380	407	434	461	488	515	542	569	596	623	650	677	704
6′	216	244	272	300	328	356	384	412	440	468	496	524	552	580	608	636	664	692	720	748
7′	241	270	299	328	357	386	415	444	473	502	531	560	589	618	647	676	705	734	763	792
8′	266	296	326	356	386	416	446	476	506	536	566	596	626	656	686	716	746	776	806	836
9′	291	322	353	384	415	446	477	508	539	570	601	632	663	694	725	756	787	818	849	880
10′	316	348	380	412	444	476	508	540	572	604	636	668	700	732	764	796	828	860	892	924
11′	341	374	407	440	473	506	539	572	605	638	671	704	737	770	803	836	869	902	935	968
12′	366	400	434	468	502	536	570	604	638	672	706	740	774	808	842	876	910	944	978	1012
13′	391	426	461	496	531	566	601	636	671	706	741	776	811	846	881	916	951	986	1021	1056
14′	416	452	488	524	560	596	632	668	704	740	776	812	848	884	920	956	992	1028	1064	1100
15′	441	478	515	552	589	626	663	700	737	774	811	848	885	922	959	996	1033	1070	1107	1144
16′	466	504	542	580	618	656	694	732	770	808	846	884	922	960	998	1036	1074	1112	1150	1188
17′	491	530	569	608	647	686	725	764	803	842	881	920	959	998	1037	1076	1115	1154	1193	1232
18′	516	556	596	636	676	716	756	796	836	876	916	956	996	1036	1076	1116	1156	1196	1236	1276
19′	541	582	623	664	705	746	787	828	869	910	951	992	1033	1074	1115	1156	1197	1238	1279	1320
20′	566	608	650	692	734	776	818	860	902	944	986	1028	1070	1112	1154	1196	1238	1280	1322	1364
21′	591	634	677	720	763	806	849	892	935	978	1021	1064	1107	1150	1193	1236	1279	1322	1365	1408
22′	616	660	704	748	792	836	880	924	968	1012	1056	1100	1144	1188	1232	1276	1320	1364	1408	1452
23′	641	686	731	776	821	866	911	956	1001	1046	1091	1136	1181	1226	1271	1316	1361	1406	1451	1496
24′	666	712	758	804	850	896	942	988	1034	1080	1126	1172	1218	1264	1310	1356	1402	1448	1494	1540

Figure 4.10 (cont.)

Rooms with 12′ Ceilings—Total Area - 4 Walls and Ceiling (in square feet)

	3′	4′	5′	6′	7′	8′	9′	10′	11′	12′	13′	14′	15′	16′	17′	18′	19′	20′	21′	22′
3′	153	180	207	234	261	288	315	342	369	396	423	450	477	504	531	558	585	612	639	666
4′	180	208	236	264	292	320	348	376	404	432	460	488	516	544	572	600	628	656	684	712
5′	207	236	265	294	323	352	381	410	439	468	497	526	555	584	613	642	671	700	729	758
6′	234	264	294	324	354	384	414	444	474	504	534	564	594	624	654	684	714	744	774	804
7′	261	292	323	354	385	416	447	478	509	540	571	602	633	664	695	726	757	788	819	850
8′	288	320	352	384	416	448	480	512	544	576	608	640	672	704	736	768	800	832	864	896
9′	315	348	381	414	447	480	513	546	579	612	645	678	711	744	777	810	843	876	909	942
10′	342	376	410	444	478	512	546	580	614	648	682	716	750	784	818	852	886	920	954	988
11′	369	404	439	474	509	544	579	614	649	684	719	754	789	824	859	894	929	964	999	1034
12′	396	432	468	504	540	576	612	648	684	720	756	792	828	864	900	936	972	1008	1044	1080
13′	423	460	497	534	571	608	645	682	719	756	793	830	867	904	941	978	1015	1052	1089	1126
14′	450	488	526	564	602	640	678	716	754	792	830	868	906	944	982	1020	1058	1096	1134	1172
15′	477	516	555	594	633	672	711	750	789	828	867	906	945	984	1023	1062	1101	1140	1179	1218
16′	504	544	584	624	664	704	744	784	824	864	904	944	984	1024	1064	1104	1144	1184	1224	1264
17′	531	572	613	654	695	736	777	818	859	900	941	982	1023	1064	1105	1146	1187	1228	1269	1310
18′	558	600	642	684	726	768	810	852	894	936	978	1020	1062	1104	1146	1188	1230	1272	1314	1356
19′	585	628	671	714	757	800	843	886	929	972	1015	1058	1101	1144	1187	1230	1273	1316	1359	1402
20′	612	656	700	744	788	832	876	920	964	1008	1052	1096	1140	1184	1228	1272	1316	1360	1404	1448
21′	639	684	729	774	819	864	909	954	999	1044	1089	1134	1179	1224	1269	1314	1359	1404	1449	1494
22′	666	712	758	804	850	896	942	988	1034	1080	1126	1172	1218	1264	1310	1356	1402	1448	1494	1540
23′	693	740	787	834	881	928	975	1022	1069	1116	1163	1210	1257	1304	1351	1398	1445	1492	1539	1586
24′	720	768	816	864	912	960	1008	1056	1104	1152	1200	1248	1296	1344	1392	1440	1488	1536	1584	1632

Square Footage for Single Floor, Ceiling or Wall Area

	3′	4′	5′	6′	7′	8′	9′	10′	11′	12′	13′	14′	15′	16′	17′	18′	19′	20′	21′	22′
3′	9	12	15	18	21	24	27	30	33	36	39	42	45	48	51	54	57	60	63	66
4′	12	16	20	24	28	32	36	40	44	48	52	56	60	64	68	72	76	80	84	88
5′	15	20	25	30	35	40	45	50	55	60	65	70	75	80	85	90	95	100	105	110
6′	18	24	30	36	42	48	54	60	66	72	78	84	90	96	102	108	114	120	126	132
7′	21	28	35	42	49	56	63	70	77	84	91	98	105	112	119	126	133	140	147	154
8′	24	32	40	48	56	64	72	80	88	96	104	112	120	128	136	144	152	160	168	176
9′	27	36	45	54	63	72	81	90	99	108	117	126	135	144	153	162	171	180	189	198
10′	30	40	50	60	70	80	90	100	110	120	130	140	150	160	170	180	190	200	210	220
11′	33	44	55	66	77	88	99	110	121	132	143	154	165	176	187	198	209	220	231	242
12′	36	48	60	72	84	96	108	120	132	144	156	168	180	192	204	216	228	240	252	264
13′	39	52	65	78	91	104	117	130	143	156	169	182	195	208	221	234	247	260	273	286
14′	42	56	70	84	98	112	126	140	154	168	182	196	210	224	238	252	266	280	294	308
15′	45	60	75	90	105	120	135	150	165	180	195	210	225	240	255	270	285	300	315	330
16′	48	64	80	96	112	128	144	160	176	192	208	224	240	256	272	288	304	320	336	352
17′	51	68	85	102	119	136	153	170	187	204	221	238	255	272	289	306	323	340	357	374
18′	54	72	90	108	126	144	162	180	198	216	234	252	270	288	306	324	342	360	378	396
19′	57	76	95	114	133	152	171	190	209	228	247	266	285	304	323	342	361	380	399	418
20′	60	80	100	120	140	160	180	200	220	240	260	280	300	320	340	360	380	400	420	440
21′	63	84	105	126	147	168	189	210	231	252	273	294	315	336	357	378	399	420	441	462
22′	66	88	110	132	154	176	198	220	242	264	286	308	330	352	374	396	418	440	462	484
23′	69	92	115	138	161	184	207	230	253	276	299	322	345	368	391	414	437	460	483	506
24′	72	96	120	144	168	192	216	240	264	288	312	336	360	384	408	432	456	480	504	528

Square Footage for Single Floor, Ceiling or Wall Area

	23′	24′	25′	26′	27′	28′	29′	30′	31′	32′	33′	34′	35′	36′	37′	38′	39′	40′	41′	42′
25′	575	600	625	650	675	700	725	750	775	800	825	850	875	900	925	950	975	1000	1025	1050
26′	598	624	650	676	702	728	754	780	806	832	858	884	910	936	962	988	1014	1040	1066	1092
27′	621	648	675	702	729	756	783	810	837	864	891	918	945	972	999	1026	1053	1080	1107	1134
28′	644	672	700	756	756	784	812	840	868	896	924	952	980	1008	1036	1064	1092	1120	1148	1176
29′	667	696	725	754	783	812	841	870	899	928	957	986	1015	1044	1073	1102	1131	1160	1189	1218
30′	690	720	750	780	810	840	870	900	930	960	990	1020	1050	1080	1110	1140	1170	1200	1230	1260
31′	713	744	775	806	837	868	899	930	961	992	1023	1054	1085	1116	1147	1178	1209	1240	1271	1302
32′	736	768	800	832	864	896	928	960	992	1024	1056	1088	1120	1152	1184	1216	1248	1280	1312	1344
33′	759	792	825	858	891	924	957	990	1023	1056	1089	1122	1155	1188	1221	1254	1287	1320	1353	1386
34′	782	816	850	884	918	952	986	1020	1054	1088	1122	1156	1190	1224	1258	1292	1326	1360	1394	1428
35′	805	840	875	910	945	980	1015	1050	1085	1120	1155	1190	1225	1260	1295	1330	1365	1400	1435	1470
36′	828	864	900	936	972	1008	1044	1080	1116	1152	1182	1224	1260	1296	1332	1368	1404	1440	1476	1512
37′	851	888	925	962	999	1036	1073	1110	1147	1184	1221	1258	1295	1332	1369	1406	1443	1480	1517	1554
38′	874	912	950	988	1026	1064	1102	1140	1178	1216	1254	1292	1330	1368	1406	1444	1482	1520	1558	1596
39′	897	936	975	1014	1053	1092	1131	1170	1209	1248	1287	1326	1365	1404	1443	1482	1521	1560	1599	1638
40′	920	960	1000	1040	1080	1120	1160	1200	1240	1280	1320	1360	1400	1440	1480	1520	1560	1600	1640	1680
41′	941	984	1025	1066	1107	1148	1189	1230	1271	1312	1353	1394	1435	1476	1517	1558	1599	1640	1681	1722
42′	966	1008	1050	1092	1134	1176	1218	1260	1302	1344	1386	1428	1470	1512	1554	1596	1638	1680	1722	1764
43′	989	1032	1075	1118	1161	1204	1247	1290	1333	1376	1419	1462	1505	1548	1591	1634	1677	1720	1763	1806
44′	1012	1056	1100	1144	1188	1232	1276	1320	1364	1408	1452	1496	1540	1584	1628	1672	1716	1760	1804	1848
45′	1035	1080	1125	1170	1215	1260	1305	1350	1395	1440	1485	1530	1575	1620	1665	1710	1755	1800	1845	1890
46′	1058	1104	1150	1196	1242	1288	1334	1380	1426	1472	1518	1564	1610	1656	1702	1748	1794	1840	1886	1932

Figure 4.10 (cont.)

Hauling Costs

A contractor can easily estimate hauling costs if he owns a suitable truck. If a truck is to be leased or rented, the local rate per day should be multiplied by the number of days needed. For example, if the cost were $200 per day for 20 days, the extended cost would be $4,000.

Rough Carpentry

Lumber is generally measured and sold by the *board foot*. The exceptions are trimwork (such as moldings and other millwork), which is sold by the lineal foot, and plywood or particle board, sold by the sheet or square foot.

A board foot is defined as one square foot of wood one inch thick. In estimates, it is symbolized by "B.F." The number of board feet in a piece of wood is determined by multiplying the width by the thickness in inches, multiplying the result by the length in feet, and then dividing that product by 12. For example, in order to find the B.F. of a stud that measures 2" x 4" x 8':

$$\frac{2 \times 4 \times 8}{12} = 5\text{-}1/3 \text{ or } 5.33 \text{ B.F.}$$

To figure the total B.F. of an entire wall of studding, follow the formula above and then multiply the number of individual studs to be used:

$$\frac{2'' \times 4'' \times 8' \times 9 \text{ studs}}{12} = 47.97 \text{ (round to 48) B.F.}$$

Published charts can be obtained which show the exact B.F. for various lengths of dimensioned lumber. A sample chart is shown in Figure 4.14. To determine how many B.F. are in a 14', 2" x 4" piece of lumber, refer to the chart for the answer, 9.33 B.F. Extensive board feet conversion charts can be purchased in most technical bookstores.

Checklist for Damage Takeoff

1. Tear-out and debris removal
2. Rough carpentry, such as studding, joists, and subflooring
3. Finish carpentry (windows, doors, mouldings, and finish flooring)
4. Plaster, drywall, sheetrock, etc.
5. Floor coverings
6. Painting, wall covering, etc.
7. Plumbing – mechanical
8. HVAC
9. Electrical
10. General cleaning
11. Miscellaneous, including exterior
12. Other

Figure 4.11

020 700	Selective Demolition	CREW	DAILY OUTPUT	MAN-HOURS	UNIT	BARE COSTS				TOTAL INCL O&P	
						MAT.	LABOR	EQUIP.	TOTAL		
720 3090	Maximum	2 Carp	300	.053	L.F.		1.14		1.14	1.83	720
3100	Ceiling trim	2 Clab	1,000	.016			.27		.27	.43	
3120	Chair rail		1,200	.013			.22		.22	.36	
3140	Railings with balusters		240	.067			1.12		1.12	1.80	
3160	Wainscoting		700	.023	S.F.		.39		.39	.62	
9000	Minimum labor/equipment charge	1 Clab	4	2	Job		34		34	54	
724 0010	PLUMBING DEMOLITION										724
1020	Fixtures, including 10′ piping										
1100	Bath tubs, cast iron	1 Plum	4	2	Ea.		49		49	75	
1120	Fiberglass		6	1.330			33		33	50	
1140	Steel		5	1.600			39		39	60	
1200	Lavatory, wall hung		10	.800			19.55		19.55	30	
1220	Counter top		8	1			24		24	37	
1300	Sink, steel or cast iron, single		8	1			24		24	37	
1320	Double		7	1.140			28		28	43	
1400	Water closet, floor mounted		8	1			24		24	37	
1420	Wall mounted		7	1.140			28		28	43	
1500	Urinal, floor mounted		4	2			49		49	75	
1520	Wall mounted		7	1.140			28		28	43	
1600	Water fountains, free standing		8	1			24		24	37	
1620	Recessed		6	1.330			33		33	50	
2000	Piping, metal, to 2″ diameter		200	.040	L.F.		.98		.98	1.50	
2050	To 4″ diameter		150	.053			1.30		1.30	1.99	
2100	To 8″ diameter	2 Plum	100	.160			3.91		3.91	6	
2150	To 16″ diameter	"	60	.267			6.50		6.50	9.95	
2240	Toilet partitions, see division 020-732										
2250	Water heater, 40 gal.	1 Plum	6	1.330	Ea.		33		33	50	
6000	Remove and reset fixtures, minimum		6	1.330			33		33	50	
6100	Maximum		4	2			49		49	75	
9000	Minimum labor/equipment charge		2	4	Job		98		98	150	
726 0010	ROOFING AND SIDING DEMOLITION										726
1000	Deck, roof, concrete plank	B-13	1,680	.033	S.F.		.61	.27	.88	1.26	
1100	Gypsum plank		3,900	.014			.26	.12	.38	.54	
1150	Metal decking		3,500	.016			.29	.13	.42	.61	
1200	Wood, boards, tongue and groove, 2″ x 6″	2 Clab	960	.017			.28		.28	.45	
1220	2″ x 10″		1,040	.015			.26		.26	.42	
1280	Standard planks, 1″ x 6″		1,080	.015			.25		.25	.40	
1320	1″ x 8″		1,160	.014			.23		.23	.37	
1340	1″ x 12″		1,200	.013			.22		.22	.36	
2000	Gutters, aluminum or wood, edge hung	1 Clab	200	.040	L.F.		.67		.67	1.08	
2100	Built-in		100	.080	"		1.35		1.35	2.16	
2500	Roof accessories, plumbing vent flashing		14	.571	Ea.		9.65		9.65	15.45	
2600	Adjustable metal chimney flashing		9	.889	"		15		15	24	
3000	Roofing, built-up, 5 ply roof, no gravel	B-2	1,600	.025	S.F.		.43		.43	.69	
3100	Gravel removal, minimum		5,000	.008			.14		.14	.22	
3120	Maximum		2,000	.020			.35		.35	.55	
3400	Roof insulation board		3,900	.010			.18		.18	.28	
4000	Shingles, asphalt strip		3,500	.011			.20		.20	.32	
4100	Slate		2,500	.016			.28		.28	.44	
4300	Wood		2,200	.018			.31		.31	.50	
4500	Skylight to 10 S.F.	1 Clab	4	2	Ea.		34		34	54	
5000	Siding, metal, horizontal		300	.027	S.F.		.45		.45	.72	
5020	Vertical		280	.029			.48		.48	.77	
5200	Wood, boards, vertical		280	.029			.48		.48	.77	
5220	Clapboards, horizontal		260	.031			.52		.52	.83	
5240	Shingles		250	.032			.54		.54	.87	
5260	Textured plywood		500	.016			.27		.27	.43	
9000	Minimum labor/equipment charge		2	4	Job		67		67	110	

For expanded coverage of these items see *Means Site Work Cost Data 1989*

19

Figure 4.12

**COST
ANALYSIS**

SHEET NO.

PROJECT

ESTIMATE NO.

ARCHITECT

DATE

TAKE OFF BY: QUANTITIES BY: PRICES BY: EXTENSIONS BY: CHECKED BY:

DESCRIPTION	SOURCE/DIMENSIONS			QUANTITY	UNIT	MATERIAL		LABOR		EQ./TOTAL	
						UNIT COST	TOTAL	UNIT COST	TOTAL	UNIT COST	TOTAL
Kitchen											
9' x 7'4" x 8' ht.											
1) Remove base cabinets											
8 LF @ .2HR = 1.6HR				1.6	HR						
2) Remove alum. windows											
1 @ .7HR				.7	HR						
3) Remove drywall walls & ceiling											
319 SF @ .008HR				2.5	HR						
Total tear out				4.8	HR						

Figure 4.13

Several years ago, most insurance companies insisted that contractors use board feet measurements for estimates involving rough carpentry. Recently, insurance companies have accepted measurements in lineal feet. Adjusters are always looking for ways to save time, and computing B.F. is time-consuming.

Rough carpentry includes framing, sheathing, roof boards, rough flooring, window and door frames, forms, scaffolding, fences, furring, and rough stairways. Since, in most cases, rough carpentry is the load-bearing support for finish carpentry items, the extent of damage to rough carpentry is generally not readily visible. The contractor should always carry a tool such as a short crowbar for probing. The following example illustrates the importance of this practice.

A small area of studded drywall partition was visibly damaged by fire. A portion of the drywall was destroyed, revealing a charred section of studding. The contractor performing a quantity takeoff cannot see how far back the studding was damaged because it is still covered by drywall. The only solution is to remove more sections of the drywall, even if undamaged, in order to fully inspect the studding. The additional damage to the drywall (removed for investigation) is considered part of the original fire loss to the structure, having been necessary to determine the extent of the damage.

The kitchen in our example required replacement of 3 (2" x 6") ceiling joists. They would be listed on the quantity takeoff as shown in Figure 4.15.

Electrical

The important factors in estimating electrical repair are specifications, materials, and labor. Most electricians provide an itemized breakdown of labor and materials. If the electrician is a subcontractor, the general contractor may incorporate the electrician's estimate into the overall estimate.

Some general contractors quantify electrical repairs on a per-circuit basis. This basis encompasses both the labor and material. Using this method, each outlet, switch, or fixture is counted. Any special circuits, and, of course, the service entrance

Board Feet per Lineal Foot in Various Sizes of Lumber						
Lumber Dimension	Lumber Length					
	8'	10'	12'	14'	16'	18'
2" x 3"	4	5	6	7	8	9
2" x 4"	5.33	6.66	8	9.33	10.66	12
2" x 6"	8	10	12	14	16	18

Formula for finding board foot per lineal foot:

$$\frac{\text{Base x Height}}{12}$$

Example: 2" x 12"

$$\frac{2 \times 12}{12} = \frac{24}{12} = 2 \text{ BF/LF of 2" x 12"}$$

Figure 4.14

Means Forms

COST ANALYSIS

SHEET NO.				
PROJECT			ESTIMATE NO.	
ARCHITECT			DATE	
TAKE OFF BY:	QUANTITIES BY:	PRICES BY:	EXTENSIONS BY:	CHECKED BY:

DESCRIPTION	SOURCE/DIMENSIONS			QUANTITY	UNIT	MATERIAL		LABOR		EQ./TOTAL	
						UNIT COST	TOTAL	UNIT COST	TOTAL	UNIT COST	TOTAL
Replace 2"x 6" ceiling joists	3#	x	10'	30	LF						

Figure 4.15

69

(if it also must be replaced) is considered in the quantity takeoff. For example, replacing a light fixture and rewiring one circuit (including two standard outlets) in our example kitchen would be written as shown in Figure 4.16.

Heating, Ventilating, Air Conditioning (HVAC)

All but minor repairs to heating and air conditioning systems should be estimated by an HVAC contractor. The general contractor should at least list the major items that are damaged, identify the materials, and discuss the probable cost of restoration with a competent HVAC contractor.

Means *Means Mechanical Cost Data* provides quantities and pricing for HVAC items. Our small example of a kitchen fire did not involve HVAC damage, but the actual case examples discussed later in this book will include this element.

Plumbing

This is another highly specialized area that is best estimated by an experienced plumbing contractor. An estimator should, however, develop a working knowledge of the plumbing trade so that he can discuss any questions with an experienced plumber to produce a takeoff which includes the materials needed and an accurate scope of repairs. (See *Means Plumbing Cost Data*, or *Fundamentals of the Construction Process*, by Kweku Bentil, R.S. Means Company, Inc., 1989, for information on plumbing systems and takeoff.)

Plaster, Drywall, and Sheetrock

Estimating quantities of plaster, including labor, depends on the type of plaster lath desired, the required thickness of the plaster, and the number of coats to be applied. Plaster is generally figured by the square yard. The area to be plastered can be obtained by dividing the number of square feet by nine to arrive at the number of square yards. For example:

27 S.F. = 3 S.Y.

For drywall or sheetrock, a takeoff should include not only amounts of wallboard material and labor, but also furring and blocking. The area to be covered is computed after deducting floor and window openings to obtain the actual number of square feet of wall and ceiling area.

In our kitchen fire example, a substantial portion of the ceiling drywall is damaged. A partial area measuring 6' x 8' must be replaced. This would be written as shown in Figure 4.17.

Finish Carpentry

Finish carpentry consists of: interior and exterior trim, doors and windows (including frames), screens, wood siding, stairs, cabinet work, and finish floors. The units of measure used in estimating finish carpentry are usually linear feet (L.F.) for trim work and some cabinetry items. Square feet should be used when estimating replacement of finished hardwood floors. When replacing a solid wood door, the quantity (in this case, 1) and size of the door (in this case, 6'-8" x 3') is enough. For our example kitchen, the finished carpentry is estimated as shown in Figure 4.18.

The complete quantity takeoff for the kitchen example would appear on one page, as shown in Figure 4.19.

COST ANALYSIS

		SHEET NO.
PROJECT		ESTIMATE NO.
ARCHITECT		DATE

TAKE OFF BY:	QUANTITIES BY:	PRICES BY:	EXTENSIONS BY:	CHECKED BY:

DESCRIPTION	SOURCE/DIMENSIONS			QUANTITY	UNIT	MATERIAL		LABOR		EQ./TOTAL	
						UNIT COST	TOTAL	UNIT COST	TOTAL	UNIT COST	TOTAL
Remove and replace one circuit incl 3 outlets, 3 switches				6							

Figure 4.16

COST ANALYSIS

SHEET NO.													
PROJECT								ESTIMATE NO.					
ARCHITECT								DATE					
TAKE OFF BY:		QUANTITIES BY:		PRICES BY:			EXTENSIONS BY:			CHECKED BY:			

DESCRIPTION	SOURCE/DIMENSIONS			QUANTITY	UNIT	MATERIAL		LABOR		EQ./TOTAL	
						UNIT COST	TOTAL	UNIT COST	TOTAL	UNIT COST	TOTAL
Replace area of drywall 1/2" Taped & finished	6'	×	8'	48	SF						

Figure 4.17

Means Forms

COST ANALYSIS

					SHEET NO.	
PROJECT					ESTIMATE NO.	
ARCHITECT					DATE	
TAKE OFF BY:	QUANTITIES BY:	PRICES BY:	EXTENSIONS BY:	CHECKED BY:		

DESCRIPTION	SOURCE/DIMENSIONS			QUANTITY	UNIT	MATERIAL		LABOR		EQ./TOTAL	
						UNIT COST	TOTAL	UNIT COST	TOTAL	UNIT COST	TOTAL
Replace solid wood door	6'8"	x	3'	1							

Figure 4.18

Means Forms

COST ANALYSIS

SHEET NO. _____

PROJECT _____ ESTIMATE NO. _____

ARCHITECT _____ DATE _____

TAKE OFF BY: _____ QUANTITIES BY: _____ PRICES BY: _____ EXTENSIONS BY: _____ CHECKED BY: _____

DESCRIPTION	SOURCE/DIMENSIONS			QUANTITY	UNIT	MATERIAL		LABOR		EQ./TOTAL	
						UNIT COST	TOTAL	UNIT COST	TOTAL	UNIT COST	TOTAL
Kitchen											
9' x 7'4" x 8' ht.											
1) Tear out and debris removal				4.8	HR						
2) replace 2" x 6" ceiling joists	(3)	X	10'	30	LF						
3) remove and replace one circuit incl. 3 outlets, 3 switches				3	HR						
				6	#						
4) replace area of drywall ½" taped & finished	6'	X	8'	48	SF						
5) replace solid wood door	6'8"	X	3'	1							

Figure 4.19

74

Pricing the Quantity Takeoff

Once the takeoff is finished, it must be "priced out." Insurance adjusters need to be able to break an estimate down into fairly explicit detail. This is because adjusters' estimates are sometimes challenged, and may need to be explained, or justified, with detailed documentation. The industry also requires this level of detail in order to prevent fraud.

Adjusters prefer to have estimates priced out by the actual unit of measure where practical. A contractor could price out the quantity takeoff with the use of a unit cost pricing guide, such as *Means Repair and Remodeling Cost Data* or Means *Building Construction Cost Data*. These annually-updated cost guides provide material, labor, crew, and equipment prices per unit of an installed building item.

The final priced out takeoff is shown in Figure 4.20. The prices in this example have been obtained from *Means Repair and Remodeling Cost Data*, 1989. For example, line 1 of this figure, tear-out and debris removal, requires a quantity of 4.8 hours of labor. This figure must be multiplied by the unit cost for tear-out and debris removal, $17.45 (obtained from *Means Repair and Remodeling Cost Data*, 1989), to produce the price for this item, $83.76.

Contractor's Overhead and Profit

The term *overhead and profit* refers only to the general contractor's overall overhead and profit. Any subcontracted items will already include provisions for overhead and profit. Adjusters generally argue against paying overhead costs to small subcontractors, especially if they are performing the only repair involved on the claim. On the other hand, adjusters routinely expect to pay general contractors for overhead and profit.

Overhead is not considered part of the actual repair job. It includes rent, heat, and power for the contractor's office or warehouse, and most any other operations expenses. Most adjusters expect to pay 10% of the subtotal of the estimate for overhead. Profit is usually about 10% of the job cost after overhead has been added in. However, this varies for regions of the country (it may be as much as 15% in some regions).

Some contractors charge a flat 20% for overhead and profit. Others calculate overhead and profit as follows.

Job cost x 10% = Overhead

(Job Cost + Overhead) x 10% = Profit

This calculation results in an actual percentage rate of 21% for overhead and profit, as shown in Figure 4.21.

Submitting the Estimate

The estimate should be typed onto company letterhead or a standard preprinted form, as shown in Figure 4.22. The contracting company name should always appear at the top of the sheet to avoid confusion.

COST ANALYSIS

SHEET NO.										
PROJECT							ESTIMATE NO.			
ARCHITECT							DATE			
TAKE OFF BY:	QUANTITIES BY:	PRICES BY:		EXTENSIONS BY:		CHECKED BY:				

DESCRIPTION	SOURCE/DIMENSIONS			QUANTITY	UNIT	MATERIAL		LABOR		EQ./TOTAL	
						UNIT COST	TOTAL	UNIT COST	TOTAL	UNIT COST	TOTAL
Kitchen 9' x 7'4" x 8' ht.											
1) Tear out and debris removal				4.8	HR					17.45	83 76
2) Replace 2"x6" ceil joists	(3)	×	10'	30	LF					3.	90
3) Remove & replace one circuit incl.				3	HR			25.	75	25.	75
3 outlets, 3 switches				6	#	7.	42				42
4) Replace area of drywall 1/2" taped & finished	6'	×	8'	48	SF					.65	31 20
5) Replace solid wood door	6'8"	×	3'	1						230.	230
Totals											551 96

Figure 4.20

COST ANALYSIS

						SHEET NO.	

PROJECT _____   ESTIMATE NO. _____

ARCHITECT _____   DATE _____

TAKE OFF BY:	QUANTITIES BY:	PRICES BY:	EXTENSIONS BY:	CHECKED BY:

DESCRIPTION	SOURCE/DIMENSIONS			QUANTITY	UNIT	MATERIAL		LABOR		EQ./TOTAL	
						UNIT COST	TOTAL	UNIT COST	TOTAL	UNIT COST	TOTAL
O&P #1 Method											
* Total from figure 4.20											551 96
add 20% O&P											110 39
Total incl O&P											662 35
O&P #2 Method											
* Total from figure 4.20											551 96
add 10% overhead											55 20
subtotal											607 16
add 10% profit											60 72
Total incl O&P											667 88

Figure 4.21

DETAILED REPAIR ESTIMATE

DATE	
PAGE	
FILE #	

CUSTOMER	INSURANCE COMPANY	LOSS CAUSE
ADDRESS	ADDRESS	FILE #
PHONE	ADJUSTER	PHONE

TAKE OFF BY: QUANTITIES BY: PRICES BY: EXTENSIONS BY: CHECKED BY:

DESCRIPTION	SOURCE/DIMENSIONS			QUANTITY	UNIT	MATERIAL		LABOR		EQ./TOTAL	
						UNIT COST	TOTAL	UNIT COST	TOTAL	UNIT COST	TOTAL
Kitchen:											
9' x 7'4" x 8' ht.											
1) Tear out and debris removal				4.8	hr					17.45	83.76
2) Replace 2" x 6" ceiling joists	(3)	x	10'	30	lf					3.	90.00
3) Remove and replace one circuit, including 3 outlets, 3 switches				3	hr			25	75.00		75.00
				6	#	7	42.00				42.00
4) Replace area of drywall 1/2" taped and finished	6'	x	8'	48	sf					.65	31.20
5) Replace solid wood door	6'8"	x	3'	1						230.	230.00
									TOTALS		$551.96

Figure 4.22

Chapter 5

HOW TO SECURE PAYMENT

Conducting business face-to-face with a private individual should be fairly straightforward. A contractor makes a proposal for a specified fee; the customer agrees to the proposal and fee, and promises to pay upon completion of the work; the work is completed, and the customer pays.

This simple scenario becomes complicated when a third party, such as a lending institution with a financial interest in the property, becomes involved. Further, when a fourth party – the insurance company – enters the picture, there is more risk involved for the contractor. Who is responsible for making payment, and when, is no longer so clear-cut. To avoid the problems associated with nonpayment, the contractor must understand the legal interests of both financial institutions and insurance companies.

The important questions to ask before starting an insurance repair job are:

1. When will the claim be paid?
2. How much of the repair bill will be paid by the insurance company and how much by the property owner?
3. Can the adjuster include the contractor's name as one of the payees on the bank draft?

Financing the Job

Most insurance repair contractors finance a job themselves and then wait for payment from either the insurance company or the policyholder. Adjusters may pay periodic "construction advances" while a contractor is performing a job, but this is usually only done for restoration projects of $20,000 or more. More often on losses of $20,000 or less, the whole claim is paid even before the restoration work begins. In such cases, the repair firm should be aware of when the insurance company paid the claimant, and should work out a payment schedule with the property owner and mortgage company (if any). Specific attention should be paid to the management of the property owner's portion of the payment, and to determining which of the two parties (policyholder or mortgagee) will hold the insurance company's draft.

Bank Drafts

Insurance claims are paid by bank draft rather than by check. The contractor should understand the legal differences between the two. Drafts are not negotiable in the same manner as checks. They are affected by the contractual relationship between insurance companies and claimants. The insurance company has more control over the negotiability of a draft in the following way. A bank in which a settlement draft is deposited will likely "hold" the draft much longer than they would a check. This control serves an additional purpose, to allow an opportunity for the detection and prevention of fraud.

A contractor should forewarn his bank representative if he will be receiving a bank draft. He should find out the correct procedures for depositing such drafts, to ensure that he has use of the funds at the earliest possible date.

Establish a Working Relationship with an Adjuster

Contractors may find adjusters somewhat indifferent to their financial concerns. In fact, some adjusters appear not at all concerned for contractors who perform work and then have trouble getting paid. Adjusters often maintain this indifference as a way of preventing any appearance of conflict of interests, in their efforts to maintain ethical standards. Clearly, it is in the contractor's best interest to take the initiative in establishing a good working relationship with the adjuster.

A "good working relationship" simply means that an adjuster is honest and "aboveboard" regarding coverage limitations, and fair and timely in the payment of a claim. There have been cases where the adjuster forgets or otherwise fails to put the contractor's name on a settlement draft. The policyholder cashes the draft and spends the money, leaving the repairman in a position where his only recourse is litigation. The repairman might think, "I'll sue or file a mechanic's lien," but litigation can be a "no win" situation. Lawsuits can take years to resolve in the courts. In some states, foreclosure of a home to satisfy a lien is prohibited. In other states, it may take a year or more to collect on a mechanic's lien. The contractor must, therefore, avoid litigation. This can be done by thorough preparation and organization of the payment process.

Preparation and Organization

Potential collection problems can be eliminated, or at least minimized, by taking precautions up front to guarantee compensation for work performed. A contractor should determine who has a financial interest in the property. Are there any judgments or liens attached? If lending institutions are involved, who will control disbursement of the insurance company proceeds?

Proper papers should be signed by the property owner authorizing the contractor to perform—according to a written **Proposal** or **Work Authorization**. Once a job is completed, a **Certificate of Satisfaction** should be signed by the property owner. These documents need not be drawn up by an attorney every time a job is to be contracted; preprinted forms can be purchased for such use. Many of the forms that appear in this chapter and throughout the book can be found in *Means Forms for Building Construction Professionals*.

Work Authorization

If a contractor intends to perform work on anyone's property, he should always obtain written permission to do so. This "permission" should be submitted to the property owner in the form of a Proposal, and signed, or accepted, by the property owner (insured).

The signed Proposal authorizes the insurance repair contractor to be on the premises performing repair work agreed upon by both parties. The wording on such a document should at least include a description of the job, with prices, and an acceptance clause stating:

> *The above prices, specifications, and conditions are satisfactory and are hereby accepted. The contractor is authorized to do the work as specified. Payment will be made as outlined above.*
>
> *Signature* _____
>
> *Date of Acceptance* _____

This clause is inserted in many standard proposal forms. Preprinted proposals, as shown in Figure 5.1, are also available. (The example in Figure 5.1 is from *Means Forms for Building Construction Professionals*.) This form includes an authorization or acceptance clause. After the form has been completed and signed by both parties, it serves as a mutually agreed upon record of the proposal and its acceptance by the owner. The following example illustrates the legal importance of drawing up a Proposal.

> *An insurance repair contractor feels it is too much trouble to get a Proposal/Acceptance form signed for small jobs, so he neglects to complete this task. Then one day, a property owner refuses to pay for repair work completed by this contractor. The repairman files suit in Magistrate Court. However, because he cannot produce a Proposal/Acceptance indicating that he had been authorized by the property owner to perform the work, he is not able to establish his case. The judge decides in favor of the property owner.*

Clearly, the contractor's loss in the example could have been prevented by the completion of a simple form. If the property owner is hesitant about signing a Proposal, the repair contractor should simply explain that his bonding company or liability insurer requires signed Proposals for all work to be performed. In addition, state laws regulating the licensing and operation of contracting firms require that Proposals be written and signed.

Although it was common practice years ago, adjusters no longer sign work authorizations under any circumstances. Nevertheless, before the adjuster includes the contractor's name on an insurance company draft, he or she may ask to see the Proposal/Acceptance form signed by the property owner.

Certificate of Satisfaction

The Certificate of Satisfaction is a form signed by the property owner approving of the work completed by the contractor. This signed form authorizes the insurance company to pay the contractor directly. If a bank draft has been issued before completion of the job, a contractor may be tempted to neglect obtaining a Certificate of Satisfaction. The draft may be in the

Means Forms

PROPOSAL

FROM:

Peter James Contracting
6611 Watson St.
Atlanta, Georgia   30136

TO: **The Essex Corporation**
4200 Terrell Mill Rd.
Marietta, Georgia

PROPOSAL NO. _____

DATE **2-1-89**

PROJECT **The Essex Building**

LOCATION **Marietta, Georgia**

CONSTRUCTION TO BEGIN **8-1-89**

COMPLETION DATE **12-1-89**

Gentlemen:

The undersigned proposes to furnish all materials and necessary equipment and perform all labor necessary to complete the following work:

See attached itemized estimate

All of the above work to be completed in a substantial and workmanlike manner

☒ for the sum of **Three Hundred Forty-one Thousand 72/100** dollars ($ **341,000.72**)

☐ to be paid for at actual cost of Labor, Materials and Equipment plus _____ percent (_____ %)

Payments to be made as follows: **Peter James Contracting**

_____ The entire amount of the contract to be paid within **30 days** after completion.

Any alteration or deviation from the plans and specifications will be executed only upon written orders for same and will be added to or deducted from the sum quoted in this contract. All additional agreements must be in writing.

The Contractor agrees to carry Workers' Compensation and Public Liability Insurance and to pay all taxes on material and labor furnished under this contract as required by Federal laws and the laws of the State in which this work is performed.

Respectfully submitted,

Contractor **Peter James Contracting**

By _____

ACCEPTANCE

You are hereby authorized to furnish all material, equipment and labor required to complete the work described in the above proposal, for which the undersigned agrees to pay the amount stated in said proposal and according to the terms thereof.

Date **2-1** 19 **89**

Ralph Jimenez
The Essex Corporation

Figure 5.1

possession of the mortgage bank and it may appear that the contractor will soon be paid. However, the Certificate of Satisfaction should *always* be obtained. There may be "snags" in releasing the funds or the property owner may change his mind and decide that he is not completely satisfied, after all. In such cases, lack of a properly signed Certificate of Satisfaction will work against a contractor.

When the settlement draft is to be paid *after* completion of repairs, the adjuster will usually want to see the Certificate of Satisfaction. This is especially true if the adjuster is not familiar with the work of a particular repair contractor.

Occasionally, a property owner may refuse to sign the Certificate of Satisfaction. If their intentions are to avoid payment and their reasons are unfounded, the contractor may enlist the adjuster's assistance in obtaining fair payment. However, keep in mind the fact that no adjuster should be perceived as the defender or protector of a contractor's rights, as this would be a conflict of interest. The adjuster can, however, at least make a telephone call to the homeowner asking for an explanation and offering to clear up any misunderstandings. A call from the adjuster can be very influential; most property owners do not want their insurance company to think that anything underhanded is being done with the settlement money. Such situations will let a contractor know which adjusters are allies and which ones are indifferent.

If a property owner still refuses to pay and there is no diplomatic way to settle the payment, then the contractor may have to resort to legal action.

A sample Certificate of Satisfaction is shown in Figure 5.2, as it would be filled in by a contractor.

Multiple Policyholders

Contractors must learn to deal with multiple policyholders if they are to perform any work on condominiums. A condominium is legally defined by different terms in almost all 50 states. It is possible that a contractor will perform work on a condominium thinking he was duly authorized to do so by the unit owner, only to find out later that the person who gave the authorization was not authorized by the condominium association. This can be a serious problem when it comes time to collect payment for services.

If the size of the job warrants, it is advisable to consult an attorney when working on condominiums. At least be sure that the Proposal is authorized and accepted by the Board of Directors or its appointed representative. Often, bylaws of a condominium association will name a member who is authorized to act on behalf of the association with regard to insurance-related matters, including loss or claim activity.

Mortgagees

A mortgagee is a financial institution that has loaned money on a particular property and, therefore, has a financial interest in that property. Every state has legislation protecting the interests of mortgagees, no matter what. For example, even if a policyholder deliberately destroys his own property in order to fraudulently collect the insurance proceeds, his portion of the claim payment may be denied, but the innocent mortgagee is always paid. If there is a mortgage on a property, that financial institution's name almost always appears on the settlement draft. The only exception is if the claim is small enough and the adjuster knows the repairs have been completed. In that case, he might leave off the name of the mortgagee to simplify the payment process.

Government agencies such as the Federal or Farmer's Housing Administration and the Department of Housing and Urban Development (H.U.D.) often act as guarantors of a loan. That is, these agencies guarantee that in the event a loan goes into default, they will protect the interests of the lending institution. The government agency may even buy the mortgage from the lending institution and then, itself, become the mortgagee on the particular property (usually when there has been a default).

Problems arise when the mortgagee refuses to endorse the settlement draft, or deposits the draft in its own account and refuses to release the funds. For example:

Certificate of Satisfaction

7-15161
FILE NUMBER

DATE 1-10-89

BU 147
POLICY NO.   CERT. NO.

Norcross, Ga.
AGENCY AT

TO The ABC Insurance Co.   Good Ins. Agency
AGENT

OF Pennsylvania

THIS IS TO CERTIFY that the loss and damage by Fire

on the 25 day of December , 19 86 , to our home

located at 1234 Peachtree Rd., Atlanta, Ga.

has been repaired/~~replaced~~ to our entire satisfaction by

Hammer Repair Company ,

the cost of such repairs/replacements to be paid to them whose receipt shall be sufficient acquittance for all claims under the above policy for loss and damage by reason of said

Geo Smith
WITNESS

Polly Holden
INSURED

Uniform Standard Form No. 3917

Figure 5.2

A contractor performs a $12,000 repair on a seemingly solvent apartment complex. The contractor was promised by the adjuster that his name would appear on the draft; and it did.

The policy listed H.U.D. as the mortgagee. Therefore, the insurance company listed H.U.D. as a payee, along with the property owner and the contractor. The contractor accepted the property owner's word that H.U.D. required that the settlement draft be endorsed by the other payees so it could be deposited. He assured the contractor that disbursement would be made at the appropriate time once inspections had been made.

After the repairs were complete, H.U.D. refused to release the funds because the apartment complex owners were in default on their mortgage. H.U.D. had the legal right to withhold the insurance settlement proceeds. Further, they advised the contractor that he should pursue litigation against the property owner. Of course, the contractor would probably have to stand in a long line of creditors, all waiting to win a judgment against the property owner.

This situation could have been avoided. The contractor should have investigated the situation at the beginning, to identify all parties with an interest in the property. Once he saw that H.U.D. was listed as a mortgagee, he might have suspected a potential problem; H.U.D. is usually only involved when there has been a default against the primary lender. The contractor should have contacted the local H.U.D. Chief of Multi-Family Loan Management to discuss the account. He would have been instructed to file a "Status of Loan" form with that department, at which time he would have been informed of the default. All of this paperwork should be processed prior to the contractor having initiated repairs. If the status of a H.U.D. loan is in good standing, that agency may even provide an authorization letter to a contractor.

A contractor is advised to approach representatives of any local government agencies that could be involved before any insurance job arises, to become familiar with the necessary procedures for involvement in a claim.

A local financial institution, such as a bank or mortgage company, has the same legal rights exercised by H.U.D. in the example above. However, because the local institution has more of a commitment to the local community, it is unlikely to resort to the practices H.U.D. used to collect their delinquent accounts, at the expense of a contractor who acted in good faith.

In any case where a mortgagee is involved, a contractor should communicate in advance with the financial institution, if its name is to appear on the settlement draft. If it happens to be an agency of the federal government, such as the H.U.D. or the F.H.A., *proceed with caution.*

Deductibles, Depreciation, and Other Penalties

As mentioned earlier, there is a good probability that certain limiting factors will be applied to a claim. The adjuster deducts for depreciation, deductibles, and coinsurance penalties. The effect of these factors is that the property owner pays for a portion of the repair job.

The insurance repair contractor should understand his own position regarding the settlement amount and convey that information to the property owner. A contractor should never "play adjuster" by trying to explain why deductions were taken, and is advised to stay out of such matters entirely. A property owner with questions about the limiting factors on his policy should be directed to his adjuster.

However, all parties involved – the property owner, the adjuster, the mortgagee, and contractor – should be aware of any limiting factors that will be applied. At the appropriate time, the contractor should follow the recommendations presented in Chapter 1 to assist the property owner in managing the financial burdens that such limits could impose.

Despite precautions, a contractor may find himself attempting to collect the balance of payment for a repair job after he has completed the work. Before proceeding with litigation, a series of simple collection letters designed to formally request payment will serve to remind the property owner of his or her obligations to the contractor.

Chapter 6
SURETY BONDING

Competition in the construction industry is ever increasing. More and more contractors are pursuing jobs of a size and type that require surety bonding of their firms. In the past, bonds were required only for public projects funded by federal or state governments. Recently, a growing number of private project owners are requiring bonds as well.

In general, insurance losses under $300,000 do not require a "surety bonded" contractor. However, those general contractors interested in repairing large insurance losses must be bonded. Bonding is necessary because a property owner who wishes to challenge the insurance company's settlement offer will begin by attacking the credibility of its general contractor (and his estimate). If the general contractor is not bonded, the insurance company will find it difficult to defend their claim, and could end up paying more for repairs using a different, bonded contractor.

What Is a Surety Bond?

A surety bond is an agreement whereby the bonding company (usually a specialty insurance company) guarantees to a property owner that a general contractor will perform according to the contract documents. Construction surety bonds protect property owners against financial losses that could result from the default of a general contractor.

Terms
The following terms are used to refer to the parties to surety bonds: the property owner is referred to as the **obligee**, the general contractor as the **principal**, and the bonding company as the **surety**. For subcontract bonds, the general contractor is the **obligee** and the subcontractor is the **principal**. This is because the general contractor must still meet his obligations in the event of a subcontractor's default. The surety bond would then protect or indemnify the general contractor for any losses resulting from the subcontractor's failure to perform.

Types of Bonds
Surety bonds can be divided into three categories: **bid bond**, **performance bond**, and **payment bond**. The bid bond assures that the contractor will actually execute the contract at the price

that was bid, and that he will provide performance and payment bonds. The performance bond is a safeguard protecting the property owner from financial loss resulting from failure or default of a general contractor to perform the job according to the terms and conditions of the contract. The payment bond assures that all expenses associated with the project, such as labor and materials, will be paid by the general contractor.

Benefits

Construction surety bonds are beneficial to all parties involved in a major insurance repair loss, from the property owner/policyholder to the insurance company responsible for payment of the loss, as well as laborers, subcontractors, and material suppliers. Surety bonds also benefit the financial institutions that hold mortgages, independent adjusters, engineers, attorneys, and risk managers.

To the **property owner**, performance and payment bonds represent assurance that a contractor will perform the work specified in the contract and that the necessary expenses (such as labor and materials) related to the job will be paid. The owner feels more secure because of the fact that the surety-bonded contractor has passed the surety's comprehensive prequalification review.

The **insurance company** responsible for paying the loss benefits because they have assurance that the work described in the bid or estimate (on which they based their settlement) can and will actually be completed for that sum. The insurance carrier wants to be sure that its settlement will be final, with no further obligation on its part.

For the **financial institution** holding the mortgage on the property, a surety bond offers a guarantee that the property on which the loan was made will be restored to its originally "valued" condition.

How to Obtain a Surety Bond

Surety bonds are sold by licensed surety companies (usually owned by or affiliated with insurance companies). These surety or bonding companies are willing to commit their assets to a contractor in exchange for a fee or premium. The rate varies, but premiums generally range from between one to five percent of the contracting company's total contract amount for a twelve-month period.

Surety companies do not usually communicate directly with contractors. The contractor must first find an agent or broker who can assist him in getting a bond through one of the bonding companies. A list of brokers specializing in surety bonding is shown in Figure 6.1. These agencies service contractors nationwide.

A list of surety companies is provided in Figure 6.2. Most of those shown offer surety bonds for contractors. While some of these companies may require that the contractor approach them with the assistance of a broker, the surety company may be willing to identify specialized brokers or agents in the contractor's locale. Many of the companies listed are willing to consider underwriting surety bonds for small contractors with revenues of $500,000 or less. Some of the firms listed actually "wear two hats"—as broker and surety.

The surety broker guides a contractor through the process of acquiring a bond and establishing a relationship with a surety company. Obtaining a bond is not like shopping for and purchasing insurance. Qualifying for a bond is more like obtaining bank credit. Like a banker, the surety company gathers

Brokers Specializing in Surety Bonding	
American Bond Underwriters P.O. Box 1623 Boston, MA 02205-1623 Phone: (617) 843-5443	Midwest Indemnity Corp. 5550 W. Touhy Ave. Skokie, IL 60077 Phone: (312) 982-9821
Bora Brokers, Inc. 6160 N. Cicero Ave. Chicago, IL 60646 Phone: (312) 736-2320	Rucker & Associates, Inc. 119 Maple St. Carrollton, GA 30117 Phone: (404) 836-0120
Brownyard Group 20 Fourth Ave. Bay Shore, NY 11706 Phone: (516) 666-5050 (800) 645-5820	Weakley & Co. 6730 LBJ Freeway, Suite 2184 Dallas, TX 75251-6054 Phone: (214) 788-0674
Carton Associates, Inc. 14 Alleyne St. Quincy, MA 02169	Contractors of American Risk Management Association c/o Gow & Hanna, Inc. 100 Maiden Lane New York, NY 10038 Phone: (212) 509-6100

Figure 6.1

Surety Companies	
Amwest Surety Insurance Co. P.O. Box 4500 Woodland Hills, CA 91367 Phone: (818) 704-1111	The Insco/DICO Group 333 Wilshire Ave. Anaheim, CA 92803-3343 Phone: (714) 999-1471
Bolton & Co. 1045 Starks Bldg. Louisville, KY 40202 Phone: (502) 582-8361	International Underwriting Group of America 4 Executive Park Dr., Suite 2314 Atlanta, GA 30329 Phone: (404) 634-8901
Capital Indemnity Corp. 4610 University Ave. Madison, WI 53705-0900 Phone: (608) 231-4450	Kigen Insurance, Inc. 154 Patchen Dr., Suite 96 Lexington, KY 1130-2444 Phone: (516) 365-7440
FG Special Risks, Inc. 4801 Woodway, Suite 285W Houston, TX 77056 Phone: (713) 993-9200	Professional Managers, Inc. 2 N. Riverside Plaza, Suite 1460 Chicago, IL 60606-2640 Phone: (312) 559-0101

Figure 6.2

and carefully analyzes a great deal of information about the construction firm and its principals before agreeing to provide bonds.

The Prequalification Process

The prequalification process can be very time-consuming. Therefore, a contractor should expect to wait several months before becoming bonded. In the interim, he should not attempt to bid on projects that require bonds, such as work representing major insurance losses.

In prequalifying a contractor, the surety will want to know that the contractor is of good character and that his experience matches the requirements of the work to be performed. The surety also investigates the financial strength of the contractor, checking into the contractor's history of paying subcontractors and suppliers promptly, and verifying that the contractor has a good working relationship with a bank and an established line of credit. In summary, in order to accept a contractor for bonding, the surety must be satisfied that the contracting company is a well-managed, profitable business that upholds commitments and performs obligations in a timely manner.

Contractor's Submittals

Before a surety company underwrites a contractor, the construction firm submits a **business plan**. A business plan is a detailed report describing the specialty or type of construction the firm engages in, how the contractor obtains work, the firm's territory of operation, and its growth and profit objectives.

The business plan should explain the organizational structure of the firm, identifying key employees and their responsibilities. This section should include the resumes of both the key employees and the company's principals.

References should be provided in the business plan, with names, addresses, and telephone numbers of previous clients. Letters of recommendation might be included in this section. Descriptions of past projects might also be included, with the name and address of each owner, the contract price, the date started and completed, and the gross profit.

A section of the construction firm's business plan should outline how the business would continue in the event of the death or disability of a key principal. One technique for managing this risk, as recommended by sureties, is to secure life insurance on key employees with the construction firm named as the beneficiary.

One other important part of the contractor's business plan is a description of the contractor's line of credit. Sureties prefer to see an unsecured line of credit that is available for short-term cash requirements, and a separate secured line of credit obtained through a long-term financing arrangement for operational equipment or real estate.

The surety company will also request access to a contracting firm's financial records for the last three to five years (if possible). These records should include a Balance Sheet, Income Statement, Statement of Changes in Financial Position, Schedules of Contracts in Progress and Contracts Completed, Schedule of

General and Administrative Expenses, and the Accountant's Opinion page (to include Explanatory Notes). Sureties prefer audited fiscal year end statements.

The surety broker/agent assists the contractor in compiling the necessary document package for prequalification, and then submits the package to a surety company for review.

The Surety Company's Position

Once the prequalification documents are submitted, the contractor may be asked to meet with personnel from the surety considering underwriting the firm. They will likely want to meet the key people in the firm, as well as review all aspects of the contractor's operations and future plans.

Once a contractor has been prequalified and thus approved, the surety retains the option of approving or rejecting specific bonds. The surety examines each specific job contract and, if found unacceptable, may decline to write the bond even though other prequalification factors are favorable. By this process, a contractor and his surety establish an understanding as to the types of projects to be undertaken.

Chapter 7

RESIDENTIAL CASE STUDY

In this chapter, we will present a case study of an insurance repair job on a residential building. This example should help to familiarize contractors with the documentation and chronology of events involved in residential insurance repair work. We will proceed – from the initial damage-producing event – to the adjuster's inspection, evaluation and report – to the contractor's involvement in the form of a site visit, quantity takeoff, and estimate. Examples are shown of the correspondence and forms that are prepared and exchanged by the various parties involved in this process.

The example project is a single family rental unit that has been substantially damaged by fire. The fire is originally reported (by the local, semi-volunteer fire department) to have started in the living room. Upon closer inspection, the insurance adjuster surmises that the fire started in the kitchen, from burning food on the stove, as evidenced by the burn patterns in the house.

The adjuster inspects the house the day after the fire, and calls in a contractor for an estimate. The contractor evaluates the site two days after the fire. Both the adjuster's and the contractor's estimates are complete seven days later. (While seven to fourteen days is a generally accepted amount of time for obtaining estimates, seven days is a more typical time period for a contractor to prepare an estimate. Some adjusters may demand the estimate sooner than seven days in the case of a fire loss, but for the most part, seven days are allowed.)

In this case, the insurance check is issued 21 days after the fire in one lump sum. The check is made payable to the owner because there is no mortgage on the house and the owner is not sure which contractor he will employ.

The anticipated repair time from the date of the fire to the date the renters could move back in is 60 days. (Adjusters generally compute the date of occupancy in terms of 30-day increments. This allows for computation of amounts for rental loss, or alternate temporary shelter for the inhabitants.)

The Adjuster's Report

The adjuster's report to his client (Figure 7.1) shows some of the steps that an adjuster must go through to handle a claim. Not all of the points in the adjuster's report are of significance to the contractor. In common practice, the contractor does not see this report. However, a contractor should be familiar with what "adjustment" is all about, and this report provides some insight into the adjuster's function.

Job Orientation

Part of the adjuster's report is devoted to job orientation. The adjuster must "orient," or familiarize, his client (the insurance company) with the site. This is done by providing measurements and a basic description of both the insured and the loss. Generally, this information is provided under three headings: *Insured*, *Risk*, and *Origin* (as shown on page 2 of the Adjuster's Report, Figure 7.1).

The Contractor's Quantity Takeoff

The first step of the contractor's estimating process is to take measurements and perform the quantity takeoff. Figure 7.2 shows the quantity takeoff for the sample residential case history. The contractor lists the demolition and the construction cost of each item to be replaced for each room, preferably on a preprinted form such as the Means Cost Analysis form, shown in Figure 7.2. For example, under the "Kitchen" heading, the contractor includes the measurements of the kitchen, the cost to tear out and remove the burned areas ("Tear-out and debris removal including haul away"), and each item to be replaced ("Spray primer/sealant on exposed framing in walls and flooring").

Quantities for each item are listed in the Quantity column of the takeoff sheet, e.g., "2 loads," "753 S.F.," etc. The contractor keeps this handwritten takeoff for his own records and presents a typewritten priced takeoff to the adjuster.

Pricing the Takeoff

After the contractor has drawn up a complete quantity takeoff, the next step is pricing the items listed to create an estimate for the adjusters. A priced takeoff for the sample project is shown in Figure 7.3.

Source of Costs

Virtually all adjusters use a pricing guide to perform their own estimates and as a basis of comparison when evaluating estimates submitted by contractors. Many adjusters use current editions of published cost guides, such as Means *Building Construction Cost Data*, *Means Repair and Remodeling Cost Data*, or *Means Open Shop Building Construction Cost Data*.

Some adjusters use price lists that have been supplied to them by local contractors. Contractors may develop these unit cost guides for their particular area, and use them as a sales tool to promote good relations with insurance adjusters. One drawback to preparing and using these do-it-yourself guides is that local costs must be updated periodically and few contractors have the time to devote to this ongoing project. Consequently, a published, annually-updated cost guide, such as Means *Building Construction Cost Data* may be a more practical option.

OGLE & CROSA

Insurance Adjusters

P.O. BOX 16307
ATLANTA, GEORGIA 30321

---

(404) 969-7722

March 15, 1989

Mrs. A. C. Vidaillet, Claims Examiner
American Fire Ins. Co.
53 Memorial Place,
Pine Mountain, GA 30000

RE: Your Claim No.: FP6042
     Your Insured : Allen Beck
     Date of Loss : 03-01-89
     Our File No. : PC-492

Dear Mrs. Vidaillet:

This supplements our acknowledgment memo dated March 2, 1989.
Upon receipt of the assignment we made immediate contact with the
insured and inspected the risk on the same day of assignment,
March 2, 1989. The following is our preliminary report.
However, we do believe we are in a position to make
recommendations with regard to settlement.

ENCLOSURES

1. Statement of Loss
2. Fire Report,
3. P. I. L. R.
4. Square Foot Appraisal Form, M.S. Service
5. Insured's Rent Register
6. Adjuster's Estimate ($26,252.60)
7. Contractor's Estimate ($27,551.50)
8. Photographs (24)

RESERVES

We had already phoned information to you regarding reserves.
Most recently, we used facsimile to update reserves.

COVERAGE

Under policy number FP 6042 you provided coverage to Allen Beck
on property situated at 102 Aycock

EXECUTIVE OFFICES: 6611 WATSON STREET   •   UNION CITY, GEORGIA 30291

Figure 7.1

Street in Newnan, Georgia. The dwelling amount was $33,500 and there was no contents coverage. Loss of Rents is covered and subject to form DP-3. A $250 deductible is applicable to this loss.

Effective dates, etc. were confirmed at your office.

INSURED

Allen Beck is the principal owner of Beck Realty, Inc. your producer on this account. It so happens that, as a sideline he dabbles in "spec" housing and rental property.

We found him to be a reasonable and conscientious individual. He cooperated with our investigation to the extent necessary. He personally expressed some embarrassment and regret over the fact that he had to be a claimant under his own agency's book of business.

RISK

This is a single story, single family rental unit of low quality but excellently maintained. We have prepared a square foot appraisal form using the Marshall-Swift Residential Cost Handbook. It showed a replacement value of $51,352. After applying the appropriate depreciation we arrived at an ACV of $41,082.

We must point out that it is our opinion that the Marshall-Swift Guide is too high in arriving at a value for this structure. We spoke to local contractors who believed they could replace the building for $21.00 to $25.00 per square foot. This would bring the replacement value more in line with the amount of insurance. Therefore, we feel no co-insurance penalty is applicable and that the amount of insurance is adequate as compared to risk.

ORIGIN

We secured a copy of the fire report prepared by the City of Newnan Fire Department. You will see that they believe the origin of the fire was the couch in the living room. We do not agree.

Figure 7.1 (cont.)

First of all, we have ruled out suspicious origin. The burn
patterns throughout the structure reflect the absence of any
liquid flamables. Furthermore, we interviewed the tenants who
were in the structure at the time of the fire and there is
nothing to indicate incendiary origin.

The evidence supports that the fire definitely started in the
north end of the dwelling where the living room and kitchen are
located. Following the path of the fire as evidenced by the
burn patterns in the living room we concluded that the fire did
not originate in the living room (where the couch was). In fact,
it is apparent that the fire entered the living room through the
kitchen doorway.

Upon examination of burn patterns found on the kitchen cabinets
and other fixtures including the ceiling, etc. we came to the
conclusion that the fire started directly over the stove.
Evidence indicates the fire started at the stove and burned
upwards then fanned outwards in a "V" shape. There was a
severely charred cast iron pot which contained charcoalized food
debris on the stove top. Due to severe destruction it was not
possible to examine the stove controls to see if one knob might
have been left on. Nevertheless, we strongly believe the fire
originated due to tenant negligence.

ADJUSTMENT

Fortunately, the fire did not burn through to the roof structure.
It was contained within the ceiling of the dwelling. However,
there was extensive fire, heat, smoke, soot, and water damage
throughout the dwelling.

The insured called in a contactor with whom he is familiar. This
is James Construction of Athens, Georgia. They prepared an
estimate of $27,551.50. They were proposing to replace much of
the drywall in the southern end of the structure as well as
ceramic tiles in both bathrooms (also located in the south end of
the structure). That is the primary difference between our
estimate and theirs. We did not allow for replacement of the
ceramic tile nor much of the drywall in the southern most
bedrooms. Our estimate came to $26,252.60 and we have reached an
agreed price with Michael's Contracting (404) 000-0000. The
insured is not sure who will do the work, as yet. Therefore, no
contractor should be listed as a payee on your settlement draft.

Figure 7.1 (cont.)

Although your policy calls for Replacement Cost, there were certain items which are exempted from Replacement Cost coverage. These are carpeting, kitchen appliances, etc.

We agreed to 21% depreciation on carpeting (21% of $2,397.46). This came to $503.46. We agreed to 12% depreciation on appliances (12% of $576.00). This came to $69.12. The total depreciation amount came to $572.58.

As you are aware, there was a $250 deductible applicable to this loss. Therefore, the total deductions from our agreed price figure came to ($572.58 plus $250.00) $822.58. When applied to the Agreed Price figure we are left with an actual Claim Payable of $25,430.02.

With regard to Loss of Rents, the insured has provided a rent ledger which shows that he collected $375 per month for rent on that unit. He has agreed to a two month "down time" in advance and wishes to settle this portion of the claim now ($750.00).

Therefore, the total amount of your claim payment will be $26,180.02 payable to Allen Beck. We have forwarded a Proof of Loss to your insured and upon return we would like to be in a position to forward the settlement draft to him. If there is any question with regard to this proposal please feel free to contact this office.

SUBROGATION

The unit was resided in by Mr. Calvin Williams and his wife, Marilyn. They have a few small children and, apart from the modest furnishings in this unit, they have no assets. A broken-down automobile appears to be permanently parked in the front yard. We feel that there is no hope of recovering any substantial return by pursuing subrogation against these individuals.

They were not insured except for a Credit Casualty policy. This policy paid off the loan balance on the furnishings in the event of total destruction. Therefore, these people have nothing and are simply not worth pursuing.

Figure 7.1 (cont.)

Page 5
Vidaillet
March 15, 1989

<u>TITLE AND ENCUMBERANCES</u>

None.

<u>RECOMMENDATIONS AND REMARKS</u>

We request that you go ahead and issue your draft for $26,180.02.
This consists of the dwelling claim payable of $25,430.02 and the
Loss of Rents portion which came to $750. Upon receipt of your
draft we will exchange same for a properly executed Proof of Loss.

We await your response.

Very truly yours,

Peter J. Crosa, AIC
OGLE & CROSA
INSURANCE ADJUSTERS

PJC:tro
Enclosures (8)

Figure 7.1 (cont.)

STATEMENT OF LOSS

	VALUE	LOSS	CLAIM

Description of Risk

1456 SF Single Family
single story dwelling
of frame structure over
crawl space. Shingle Roof. $41,082.00

Circumstances of Loss

Fire of accidental origin
caused extensive damage by
heat, smoke, soot, water.

Loss as Determined

Contractor's Est. $27,551.50
Adjuster's Est. $26,252.60 $26,252.60

Depreciation as follows;
Carpeting - 21% of $2,397.46
    or $503.46
Kitchen Appliances - 12% of
    $576.00 or $69.12

Total Depreciation $572.58

Summary

Loss   - $26,252.60
Depreciation - $ 572.58
Deductible   - $ 250.00
                 $25,430.02 $25,430.02

	VALUE	LOSS	CLAIM
TOTALS	$41,082.00	$26,252.60	$25,430.02

Figure 7.1 (cont.)

Means Forms

COST ANALYSIS

SHEET NO. _1_

PROJECT **Allen Beck, 102 Aycock St., Newnan, Ga.**

ESTIMATE NO.

ARCHITECT

DATE **3·2·89**

TAKE OFF BY: **Crosa** QUANTITIES BY: PRICES BY: **Mean's** EXTENSIONS BY: CHECKED BY:

DESCRIPTION	SOURCE/DIMENSIONS			QUANTITY	UNIT	MATERIAL			LABOR			EQ./TOTAL		
						UNIT COST	TOTAL		UNIT COST	TOTAL		UNIT COST	TOTAL	
Kitchen														
11'9" x 14'4" x 8' ht.														
1) tear out & debris removal incl. haul away				2	loads									
				24	HR									
2) spray primer/sealant on exposed framing in walls & flooring				753	SF									
3) replace insulation in ceiling & 2 walls				377	SF									
4) replace drywall complete				586	SF									
5) replace moulding incl. paint-1 coat with primer				34	LF									
6) replace stdrd. grade cabinetry as follows:														
wall				19	LF									
base				9	LF									
countertop				9	LF									
backsplash				9	LF									
includes finish														
7) replace enameled dbl. sink incl. min. plumbing labor				1										
8) replace 42" hood vented				1										
9) replace free standing range 30" elec standard/white				1										
10) replace lite fixture -flourescent 24"				1										
11) replace dbl. hung wood windows incl. screen & paint	32"	x	28"	2										
12) paint walls & ceil.				586	SF									
13) replace 12" acoustical tile in kitchen ceil.				168	SF									
14) replace vinyl flooring				20	SY									

Figure 7.2

Means Forms

COST ANALYSIS

	SHEET NO. **2**
PROJECT **Beck**	ESTIMATE NO.
ARCHITECT	DATE **3-2-89**

TAKE OFF BY: **Crosa** QUANTITIES BY: PRICES BY: **Mean's** EXTENSIONS BY: CHECKED BY:

DESCRIPTION	SOURCE/DIMENSIONS			QUANTITY	UNIT	MATERIAL		LABOR		EQ./TOTAL	
						UNIT COST	TOTAL	UNIT COST	TOTAL	UNIT COST	TOTAL
15) replace solid wood exterior door	3'	×	6'8"	1							
16) replace screen door (add'l. paint allowed in #12 above)				1							
Utility Room											
9' × 11'9" × 8' ht.											
17) tear out and debris removal				8	HR						
— haul away incl. in #1											
18) spray primer/seal on framing in walls & flooring				543	SF						
19) replace insulation in ceil. & one wall				178	SF						
20) replace drywall complete				438	SF						
21) replace moulding incl. paint				40	LF						
22) replace 52 gal. water heater-elec.				1							
23) replace pine shelving 1"×9" incl. paint	1"	×	9"	14	LF						
24) replace exhaust fan 16"				1							
25) paint walls & ceiling				438	SF						
26) replace vinyl flooring				12	SY						
27) replace hollow core door & hardware				1							
28) paint #27 above				2	sides						

Figure 7.2 (cont.)

Means Forms

COST ANALYSIS

SHEET NO. **3**

PROJECT **Beck**

ESTIMATE NO.

ARCHITECT

DATE **3·2·89**

TAKE OFF BY: **Crosa**   QUANTITIES BY: PRICES BY: **Mean's**   EXTENSIONS BY: CHECKED BY:

DESCRIPTION	SOURCE/DIMENSIONS			QUANTITY	UNIT	MATERIAL		LABOR		EQ./TOTAL	
						UNIT COST	TOTAL	UNIT COST	TOTAL	UNIT COST	TOTAL
Living Room											
13'9" X 17'4" X 8'ht.											
29) tear out & debris removal				8	HR						
– haul away				1	load						
30) spray primer/seal on framing				974	SF						
31) replace insulation in ceil. & two walls				487	SF						
32) replace drywall– walls & ceiling				735	SF						
33) replace moulding				62	LF						
34) replace ext. door	3'	X	6'8"	1							
35) replace screen door				1							
36) replace glass in door lites – two	12"	X	60"	10	SF						
– incl. labor to install & repair mldgs.				3	HR						
37) replace dbl. hung wood windows incl. paint.	40"	X	48"	2							
38) paint walls & ceilings				735	SF						
39) replace carpet & pad incl. installation				37.33	SY						
Hall											
3' X 17'4" X 8 ht.											
40) tear out & debris removal				12	HR						
41) spray primer/sealant				429	SF						
42) replace insulation in ceiling				52	SF						
43) replace drywall				377	SF						
44) replace mldg. incl. paint				41	LF						
45) replace hollow core door				4							
46) paint doors per side				8							

Figure 7.2 (cont.)

103

**COST
ANALYSIS**

SHEET NO. **4**

PROJECT **Beck**

ESTIMATE NO.

ARCHITECT

DATE **3-2-89**

TAKE OFF BY: **Crosa** QUANTITIES BY: PRICES BY: **Mean's**   EXTENSIONS BY: CHECKED BY:

DESCRIPTION	SOURCE/DIMENSIONS			QUANTITY	UNIT	MATERIAL		LABOR		EQ./TOTAL	
						UNIT COST	TOTAL	UNIT COST	TOTAL	UNIT COST	TOTAL
47) replace smoke detector				1							
48) replace door bell				1							
49) paint walls & ceil.				377	SF						
50) replace carpet pad incl. installation				10.66	SY						
Front Bedroom incl. closet											
10' x 13' 9" x 8' ht.											
51) tear out & debris removal				12	HR						
52) primer/sealant on walls & ceiling framing				655	SF						
53) replace insulation in ceiling & one wall				248	SF						
54) replace drywall				518	SF						
55) replace moulding				48	LF						
56) replace dbl. hung wood windows incl. paint	40"	X	48"	2							
57) paint walls & ceil.				518	SF						
58) replace carpet & pad				26.66	SY						
Hall Bath											
5' x 7' x 8' ht.											
59) tear out & debris removal				4	HR						
—1 load haul away				1	load						
60) primer/seal on ceiling frame only				35	SF						
61) replace insulation				35	SF						
62) replace drywall in ceiling				35	SF						
63) paint moulding				24	LF						
64) replace exhaust fan 24" round				1							
65) paint ceiling & partial wall				147	SF						

Figure 7.2 (cont.)

Means Forms

COST ANALYSIS

SHEET NO. **5**

PROJECT **Beck**

ESTIMATE NO.

ARCHITECT

DATE **3-2-89**

TAKE OFF BY: **Crosa** QUANTITIES BY: PRICES BY: **Mean's** EXTENSIONS BY: CHECKED BY:

DESCRIPTION	SOURCE/DIMENSIONS			QUANTITY	UNIT	MATERIAL		LABOR		EQ./TOTAL	
						UNIT COST	TOTAL	UNIT COST	TOTAL	UNIT COST	TOTAL
66) repair ceramic tile floor – min. charge				min.							
67) chem. clean floor and all fixtures				16	HR						
68) clean & regrout ceramic tiles in floor & enclosure				139	SF						
Rear Bedroom incl. closet											
16' × 11' 9" × 8' ht.											
69) tear out & debris removal				2	HR						
70) clean & seal floor				188	SF						
71) clean & seal walls & ceiling				632	SF						
72) paint walls & ceil.				632	SF						
73) paint moulding				56	LF						
74) replace windows – dbl. hung wood incl. paint	40"	×	48"	2							
75) paint pine shelving				48	LF						
76) replace carpet & pad				21.33	SY						
77) replace drop ceil. 2' tiles - acoustical				188	SF						
Hall Closet											
3' 2" × 2' × 8' ht.											
78) clean & seal floor, walls, ceil.				95	SF						
79) paint walls & ceil.				89	SF						
80) paint shelves				12	LF						
81) replace carpet & pad				1	SY						
82) paint moulding				10	LF						

Figure 7.2 (cont.)

COST ANALYSIS

SHEET NO. **6**

PROJECT **Beck**

ESTIMATE NO.

ARCHITECT

DATE **3-2-89**

TAKE OFF BY: **Crosa** QUANTITIES BY: PRICES BY: **Mean's** EXTENSIONS BY: CHECKED BY:

DESCRIPTION	SOURCE/DIMENSIONS			QUANTITY	UNIT	MATERIAL		LABOR		EQ./TOTAL	
						UNIT COST	TOTAL	UNIT COST	TOTAL	UNIT COST	TOTAL
Master Bedroom											
13' 9" x 14' x 8' ht.											
83) tear out & debris removal				2	HR						
84) clean & seal floor walls & ceiling				829	SF						
85) paint walls & ceil.				637	SF						
86) paint moulding				56	LF						
87) replace. dbl. hung wood windows as follows:	40"	X	48"	1							
	40"	X	36"	1							
88) replace carpet & pad				36.66	SY						
89) paint shelving				16	LF						
Master Bath											
5' x 7' x 8' ht.											
90) tear out & debris removal				2	HR						
91) clean & seal walls, ceiling & floor				262	SF						
92) paint walls & ceil.				147	SF						
93) paint mldgs				24	LF						
94) clean & regrout ceramic tile				139	SF						
95) chem. clean floor & all fixtures				16	HR						
96) replace heat lamp											
97) replace medicine cabinets											
98) paint shelves				4	LF						
99) replace dbl. hung wood windows incl. paint	40"	X	38"	1							
100) paint door per side				2							
101) replace flouresc. lite			48"	1							

Figure 7.2 (cont.)

Means Forms
COST ANALYSIS

PROJECT: Beck
SHEET NO. 7
ESTIMATE NO.
ARCHITECT
DATE 3-2-89
TAKE OFF BY: Crosa
QUANTITIES BY:
PRICES BY: Mean's
EXTENSIONS BY:
CHECKED BY:

DESCRIPTION	SOURCE/DIMENSIONS			QUANTITY	UNIT	MATERIAL		LABOR		EQ./TOTAL	
						UNIT COST	TOTAL	UNIT COST	TOTAL	UNIT COST	TOTAL
Miscellaneous – Mechanical & Exterior											
102) clean duct system per vent - HVAC				10							
103) replace registers – HVAC				10							
104) furnace service check				2	HR						
105) rewire electrical complete incl.											
–service entrance (200 Amp - 20 cir) (service check)				2	HR						
–9 circuits –110				24	HR						
–3 circuits - 220				16	HR						
–outlets, switches, & fixtures				16							
106) cornice repairs				24	LF						
107) siding repairs				200	SF						
108) repair roof deck				64	SF						
109) repair shingles				200	SF						
110) paint exterior siding				134	SF						
111) paint cornice				180	LF						
112) pressure clean deck & siding				10	HR						
113) permit											
114) temporary power				3	HR						

Figure 7.2 (cont.)

107

COST ANALYSIS

SHEET NO. *1*

PROJECT *Allen Beck, 102 Aycock St., Newnan, Ga.*

ESTIMATE NO.

ARCHITECT

DATE *3-2-89*

TAKE OFF BY: *Crosa* QUANTITIES BY: PRICES BY: *Mean's* EXTENSIONS BY: CHECKED BY:

DESCRIPTION	SOURCE/DIMENSIONS			QUANTITY	UNIT	MATERIAL		LABOR		EQ./TOTAL	
						UNIT COST	TOTAL	UNIT COST	TOTAL	UNIT COST	TOTAL
Kitchen											
11'9" x 14'4" x 8 ht.											
1) tear out & debris				2	loads					122	244
removal incl. haul				24	HR	17.45				17.45	418 80
away											
2) spray primer/seal. on											
exposed framing in											
walls & flooring				753	SF					.17	128 01
3) replace insulation in											
ceiling & 2 walls				377	SF					.38	143 26
4) replace drywall complete				586	SF					.65	380 90
5) replace moulding incl.											
paint-1 coat with primer				34	LF					1.62	55 08
6) replace standard											
grade cabinetry as											
follows: wall				19	LF					78.	1482 00
base				9	LF					78.	702
counter top				9	LF					78.	702
back splash				9	LF					78.	702
includes finish											
7) replace enameled											
dbl. sink incl. min.											
plumbing labor				1	ea.					310	310
8) replace 42" hood vented				1	ea.					215	215
9) replace free											
standing range											
30" elec. stdrd./white				1	ea.					315	315
10) replace lite. fixture											
flourescent 24"				1	ea.					93	93
11) replace dbl. hung											
wood windows incl.											
screen & paint	32"	x	28"	2	ea.					105	210
12) paint walls & ceil.				586	SF					.25	146 50
13) replace 12"											
acoustical tile in											
kitchen ceiling				168	SF					1.70	285 60
14) replace vinyl											
flooring				180	SF					1.83	329 40

Page Total 6,862.55

Figure 7.3

COST ANALYSIS

SHEET NO. **2**

PROJECT **Beck**

ESTIMATE NO.

ARCHITECT

DATE **3-2-89**

TAKE OFF BY: **Crosa** QUANTITIES BY: PRICES BY: **Mean's** EXTENSIONS BY: CHECKED BY:

DESCRIPTION	SOURCE/DIMENSIONS			QUANTITY	UNIT	MATERIAL		LABOR		EQ./TOTAL	
						UNIT COST	TOTAL	UNIT COST	TOTAL	UNIT COST	TOTAL
15) replace solid wood exterior door	3'	×	6'8"	1	ea.					220	220
16) replace screen door (add'l. paint allowed in #12 above)				1	ea.					96	96
Utility Room 9' x 11'9" x 8' ht.											
17) tear out and debris removal haul away incl. in #1				8	HR.					17.45	139 60
18) spray primer/seal on framing in walls & flooring				543	SF					.17	92 31
19) replace insulation in ceil. & one wall				178	SF					.38	67 64
20) replace drywall complete				438	SF					.65	284 70
21) replace moulding incl. paint				40	LF					1.62	64 80
22) replace 50 gal. water heater-elec.				1	ea.					275.	275
23) replace pine shelving 1"x 9" incl. paint	1"	×	9"	14	LF					2.51	35 14
24) replace exhaust fan 16"				1	ea.					130	130
25) paint walls & ceil.				438	SF					.25	109 50
26) replace vinyl flooring				108	SF					1.83	197 64
27) replace hollow core door & hardware				1	ea.					150	150
28) paint #27 above				2	SIDES					14.35	28 70

Page Total 1,891.03

Figure 7.3 (cont.)

COST ANALYSIS

PROJECT **Beck**			SHEET NO. **3**	
ARCHITECT			ESTIMATE NO.	
			DATE **3-2-89**	
TAKE OFF BY: **Crosa** QUANTITIES BY:	PRICES BY: **Means**	EXTENSIONS BY:	CHECKED BY:	

DESCRIPTION	SOURCE/DIMENSIONS			QUANTITY	UNIT	MATERIAL UNIT COST	MATERIAL TOTAL	LABOR UNIT COST	LABOR TOTAL	EQ./TOTAL UNIT COST	EQ./TOTAL TOTAL
Living Room											
13'9" x 17'4" x 8' ht.											
29) tear out & debris											
removal				8	HR					17.45	139 60
– haul away				1	load					122	122
30) spray primer/seal											
on framing				974	SF					.17	165 58
31) replace insulation											
in ceiling & two walls				487	SF					.38	185 06
32) replace drywall –											
walls & ceiling				735	SF					.65	477 75
33) replace moulding				62	LF					1.62	100 44
34) replace ext. door	3'	x	6'8"	1	ea.					220	2 20
35) replace screen door				1	ea.					96	96
36) replace glass in											
door lites – two	12"	x	60"	10	SF					5.95	59 50
– incl. labor to install											
& repair mldgs.				3	HR					22.45	67 35
37) replace dbl. hung											
wood window incl.											
paint	40"	x	48"	2						130	260
38) paint walls & ceilings				735	SF					.25	183 75
39) replace carpet & pad											
incl. installation				37.33	SY					14.95	558 08
Hall											
3' x 17'4" x 8' ht.											
40) tear out & debris											
removal				12	HR					17.45	209 40
41) spray primer/sealant				429	SF					.17	72 93
42) replace insulation											
in ceiling				52	SF					.38	19 76
43) replace drywall				377	SF					.65	245 05
44) replace moulding											
incl. paint				41	LF					1.62	66 42
45) replace hollow											
core door				4	ea.					150.	600
46) paint doors											
per side				8						14.35	114 80
									Page Total		3,963.47

Figure 7.3 (cont.)

COST ANALYSIS

SHEET NO. **4**

PROJECT **Beck**

ESTIMATE NO.

ARCHITECT

DATE **3·2·89**

TAKE OFF BY: **Crosa** QUANTITIES BY: PRICES BY: **Mean's** EXTENSIONS BY: CHECKED BY:

DESCRIPTION	SOURCE/DIMENSIONS			QUANTITY	UNIT	MATERIAL		LABOR		EQ./TOTAL	
						UNIT COST	TOTAL	UNIT COST	TOTAL	UNIT COST	TOTAL
47) replace smoke detector				1						15.	15
48) replace door bell				1						25.	25
49) paint walls & ceil.				377	SF					.25	94 25
50) replace carpet pad incl. installation				10.66	SY					14.95	159 37
Front Bedroom incl. closet											
10' × 13'9" × 8' ht.											
51) tear out & debris removal				12	HR					17.45	209 40
52) primer/sealant on walls & ceiling framing				655	SF					.17	111 35
53) replace insulation in ceiling & one wall				248	SF					.38	94 24
54) replace drywall				518	SF					.65	336 70
55) replace moulding				48	LF					1.62	77 76
56) replace dbl. hung wood windows incl. paint	40"	×	48"	2						130	260
57) paint walls & ceil.				518	SF					.25	129 50
58) replace carpet & pad				26.66	SY					14.95	398 57
Hall Bath											
5' × 7' × 8' ht.											
59) tear out & debris removal				4	HR					17.45	69 80
−1 load haul away				1	load					122	122
60) primer/seal on ceiling frame only				35	SF					.17	5 95
61) replace insulation				35	SF					.38	13 30
62) replace drywall in ceiling				35	SF					.65	22 75
63) paint moulding				24	LF					.21	5 04
64) replace exhaust fan 24" round				1						155	155
65) paint ceiling & partial wall				147	SF					.25	36 75

Page Total 2,341.73

Figure 7.3 (cont.)

COST ANALYSIS

SHEET NO. **5**

PROJECT **Beck**

ESTIMATE NO.

ARCHITECT

DATE **3-2-89**

TAKE OFF BY: **Crosa**   QUANTITIES BY: PRICES BY: **Mean's**   EXTENSIONS BY: CHECKED BY:

DESCRIPTION	SOURCE/DIMENSIONS			QUANTITY	UNIT	MATERIAL		LABOR		EQ./TOTAL	
						UNIT COST	TOTAL	UNIT COST	TOTAL	UNIT COST	TOTAL
66) repair ceramic tile floor - min. charge				min.						75	75
67) chem. clean floor and all fixtures				16	HR					17.45	279 20
68) clean & regrout ceramic tiles in floor & enclosure				139	SF					.82	113 98
Rear Bedroom incl. closet											
16' × 11' 9" × 8'											
69) tear out & debris removal				2	HR					17.45	34 90
70) clean & seal floor				188	SF					.20	37 60
71) clean & seal walls & ceiling				632	SF					.15	94 80
72) paint walls & ceil.				632	SF					.25	158
73) paint moulding				56	LF					.21	11 76
74) replace windows - dbl. hung wood incl. paint	40"	×	48"	2						130	260
75) paint pine shelving				48	LF					.41	19 68
76) replace carpet & pad				21.33	SY					14.95	318 88
77) replace drop ceil. 2' tiles - acoustical				188	SF					1.78	334 64
Hall Closet											
3' 2" × 2' × 8 ht.											
78) clean & seal floor, walls & ceil.				95	SF					.15	14 25
79) paint walls & ceil.				89	SF					.25	22 25
80) paint shelves				12	LF					.41	4 92
81) replace carpet & pad				1	SY					14.95	14 95
82) paint moulding				10	LF					.21	2 10

Page Total   1,796.91

Figure 7.3 (cont.)

Means Forms

COST ANALYSIS

SHEET NO. **6**

PROJECT **Beck**

ESTIMATE NO.

ARCHITECT

DATE **3·2·89**

TAKE OFF BY: **Crosa** QUANTITIES BY: PRICES BY: **Mean's** EXTENSIONS BY: CHECKED BY:

DESCRIPTION	SOURCE/DIMENSIONS			QUANTITY	UNIT	MATERIAL		LABOR		EQ./TOTAL	
						UNIT COST	TOTAL	UNIT COST	TOTAL	UNIT COST	TOTAL
<u>Master Bedroom</u>											
13'9" X 14' X 8' ht.											
83) tear out and debris removal				2	HR					17.45	34 90
84) clean & seal floor walls & ceiling				829	SF					.15	124 35
85) paint walls & ceil.				637	SF					.25	159 25
86) paint moulding				56	LF					.21	11 76
87) replace dbl. hung wood windows as follows:	40"	X	48"	1						130	130
	40"	X	36"	1						130	130
88) replace carpet & pad				36.66	SY					14.95	5 98 07
89) paint shelving				16	LF					.41	6 56
<u>Master Bath</u>											
5' X 7' X 8' ht.											
90) tear out & debris removal				2	HR					17.45	34 90
91) clean & seal walls, ceiling & floor				262	SF					.15	39 30
92) paint walls & ceil.				147	SF					.25	36 75
93) paint mldgs.				24	LF					.21	5 04
94) clean & regrout ceramic tile				139	SF					.82	113 98
95) chem-clean floor & all fixtures				16	HR					17.45	279 20
96) replace heat lamp											
97) replace medicine cabinet											
98) paint shelves				4	LF					.41	1 64
99) replace dbl. hung wood windows incl. paint	40"	X	38"	1						130.	1 30
100) paint door per side				2						14.35	28 70
101) replace flouresc. lite		48"		1						84.	84
										Page Total	1,898.40

Figure 7.3 (cont.)

COST ANALYSIS

PROJECT **Beck**

ESTIMATE NO.

ARCHITECT

DATE

TAKE OFF BY: **Crosa**   QUANTITIES BY: PRICES BY: **Mean's**   EXTENSIONS BY: CHECKED BY:

DESCRIPTION	SOURCE/DIMENSIONS				QUANTITY	UNIT	MATERIAL		LABOR		EQ./TOTAL	
							UNIT COST	TOTAL	UNIT COST	TOTAL	UNIT COST	TOTAL
Miscellaneous — Mechanical & Exterior												
102) Clean duct system per vent - HVAC					10						35.	3 50
103) Replace registers – HVAC					10						15.	1 50
104) Furnace service Check					2	HR					25.15	50 30
105) Rewire electrical complete incl.												
– service entrance												
(200 amp. – 20 cir.)												
(service check)					2	HR					24.50	49
– 9 circuits – 110					24	HR					24.50	5 88
– 3 circuits – 220					16	HR					24.50	3 92
– outlets, switches, & fixtures					16		6	96				96
106) cornice repairs					24	LF					8	1 92
107) siding repairs					200	SF					1.50	3 00
108) repair roof deck					64	SF					.85	54 40
109) repair shingles					200	SF					.45	90
110) paint exterior siding					134	SF					.32	42 88
111) paint cornice					180	LF					2.	3 60
112) pressure clean deck & siding					10	HR					24.	2 40
113) permit											95.	95
114) temporary power					3	HR					24.50	73 50
page total												3 123 08
All pages total												21 567 07
(20%) contractor's overhead & profit												4 313 41
Total estimate												25 880 48

Figure 7.3 (cont.)

The contractor can and should use prices from either published cost guides or his own company's historic cost data file in preparing estimates for adjusters. The firm's own data (based on experience) should be a source of the most accurate unit prices for that individual contractor.

An an example, Item #2 of Figure 7.3, "Spray primer/sealant on exposed framing in walls of flooring," must be applied to 753 S.F. of walls and flooring. A unit cost of .17 per S.F. is derived from p. 222 of *Means Open Shop Building Construction Cost Data*, 1989, line No. 099-220-8950 (shown in Figure 7.4). This per S.F. cost, .17, is multiplied by 753 S.F. to arrive at the total cost for this item, $128.01, which is entered in the "Total" column of Figure 7.3.

This quantity takeoff was priced using the *Means Open Shop Building Construction Cost Data*, 1989, because the site was in a rural region of Georgia where high union wages are virtually nonexistent. Labor costs will be considerably less when not using union prices. *Means Open Shop Building Construction Cost Data* is useful in such situations.

Contractor's Final Estimate

Most insurance adjusters require a typewritten quantity takeoff and estimate. Accordingly, the contractor for this job must prepare a detailed unit cost estimate showing quantities of materials and room dimensions. This estimate is typed in an orderly format on a preprinted form, as shown in Figure 7.5.

099 200 | Interior Painting

			CREW	DAILY OUTPUT	MAN-HOURS	UNIT	MAT.	LABOR	EQUIP.	TOTAL	TOTAL INCL O&P	
220	3920	Paint 2 coats, brushwork	1 Pord	220	.036	Ea.	.06	.47		.53	.83	**220**
	3940	Spray	"	250	.032	"	.06	.42		.48	.74	
	4200	Gutters and downspouts, oil base, primer coat, brushwork	2 Pord	650	.024	L.F.	.08	.32		.40	.61	
	4300	Paint 2 coats, brushwork		325	.049		.16	.64		.80	1.21	
	5000	Pipe, to 4" diameter, primer or sealer coat, oil base, brushwork		800	.020		.08	.26		.34	.51	
	5100	Spray		1,100	.014		.08	.19		.27	.39	
	5350	Paint 2 coats, brushwork		440	.036		.14	.47		.61	.92	
	5400	Spray		550	.029		.14	.38		.52	.77	
	6300	To 16" diameter, primer or sealer coat, brushwork		192	.083		.16	1.09		1.25	1.93	
	6350	Spray		240	.066		.16	.87		1.03	1.58	
	6500	Paint 2 coats, brushwork		100	.160		.33	2.09		2.42	3.73	
	6550	Spray		130	.123		.33	1.61		1.94	2.95	
	7000	Trim , wood, incl. puttying, under 6" wide										
	7200	Primer coat, oil base, brushwork	1 Pord	900	.008	L.F.	.02	.12		.14	.21	
	7250	Paint, 1 coat, brushwork		875	.009		.03	.12		.15	.23	
	7450	3 coats		370	.021		.08	.28		.36	.54	
	7470											
	7500	Over 6" wide, primer coat, brushwork	1 Pord	600	.013	L.F.	.03	.17		.20	.31	
	7550	Paint, 1 coat, brushwork		450	.017		.04	.23		.27	.42	
	7650	3 coats		190	.042		.11	.55		.66	1.01	
	8000	Cornice, simple design, primer coat, oil base, brushwork		275	.029	S.F.	.04	.38		.42	.66	
	8250	Paint, 1 coat		250	.032		.05	.42		.47	.73	
	8350	Ornate design, primer coat		150	.053		.04	.70		.74	1.17	
	8400	Paint, 1 coat		140	.057		.05	.75		.80	1.26	
	8600	Balustrades, per side, primer coat, oil base, brushwork		300	.026		.04	.35		.39	.60	
	8650	Paint, 1 coat		285	.028		.05	.37		.42	.65	
	8900	Trusses and wood frames, primer coat, oil base, brushwork		800	.010		.04	.13		.17	.25	
	8950	Spray		1,200	.006		.04	.09		.13	.18	
	9220	Paint 2 coats, brushwork		500	.016		.09	.21		.30	.44	
	9240	Spray		600	.013		.09	.17		.26	.38	
	9260	Stain, brushwork, wipe off		600	.013		.04	.17		.21	.32	
	9280	Varnish, 3 coats, brushwork		275	.029		.15	.38		.53	.78	
	9350	For latex paint, deduct					10%					
224	0010	**WALL AND CEILINGS**										**224**
	0020	Labor cost includes protection of adjacent items not painted										
	0100	Concrete, dry wall or plaster, oil base, primer or sealer coat										
	0200	Smooth finish, brushwork	1 Pord	1,900	.004	S.F.	.04	.05		.09	.13	
	0240	Roller		2,200	.003		.04	.05		.09	.12	
	0300	Sand finish, brushwork		1,700	.004		.05	.06		.11	.15	
	0340	Roller		2,100	.003		.05	.05		.10	.14	
	0380	Spray		3,750	.002		.05	.03		.08	.10	
	0800	Paint 2 coats, smooth finish, brushwork		975	.008		.08	.11		.19	.26	
	0840	Roller		1,125	.007		.09	.09		.18	.25	
	0880	Spray		2,250	.003		.11	.05		.16	.20	
	0900	Sand finish, brushwork		825	.009		.10	.13		.23	.31	
	0940	Roller		1,050	.007		.11	.10		.21	.28	
	0980	Spray		2,250	.003		.13	.05		.18	.22	
	1500											
	1600	Glaze coating, 5 coats, spray, clear	1 Pord	900	.008	S.F.	.50	.12		.62	.74	
	1640	Multicolor	"	900	.008		.60	.12		.72	.85	
	1700	For latex paint, deduct					10%					
	1800	For ceiling installations, add						25%				
	1900											
	2000	Masonry or concrete block, oil base, primer or sealer coat										
	2100	Smooth finish, brushwork	1 Pord	1,725	.004	S.F.	.04	.06		.10	.14	
	2180	Spray		3,750	.002		.06	.03		.09	.11	
	2200	Sand finish, brushwork		1,400	.005		.05	.07		.12	.18	

223

Figure 7.4

116

			DATE	3-2-89
			PAGE	1 of 5
DETAILED REPAIR ESTIMATE			FILE #	

CUSTOMER	Allen Beck	INSURANCE COMPANY		LOSS CAUSE	
ADDRESS	102 Aycock St.	ADDRESS		FILE #	
	Newnan, GA	ADJUSTER		PHONE	

DESCRIPTION	DIMENSIONS			QUANTITY	UNIT	UNIT COST	TOTAL COST
Living Room:	14'	17'	8'				
Sheetrock Walls and Ceiling				734	sf	.64	469.76
Paint Walls and Ceiling				734	sf	.24	176.16
Exterior Door	2'	x	6'8"R	1		235.00	235.00
Screen Door	3'	x	6'8"	1		90.00	90.00
Paint Doors				2		30.00	60.00
Fixed Window	1'	x	5'2"	2		120.00	240.00
Double Hung Window	3'4"	x	4'2"	1		260.00	260.00
Paint Windows (2 sides)				2		30.00	60.00
Paint Screens				2		26.00	52.00
Base Mold				62	lf	.97	60.14
Paint Base Mold	3'4"	x	4'2"	62	lf	.26	16.12
Case Opening				1		65.00	65.00
Carpet				45	yd	14.95	672.75
Underlayment				237	sf	.75	177.75
Vents - Heat				2		14.00	28.00
Hall:	3.17	17.33	8				
Sheetrock Walls and Ceiling				382.93	sf	.64	245.08
Paint Walls and Ceiling				382.93	sf	.24	91.90
Base Mold				41	lf	.97	39.77
Paint Base Mold				41	lf	.26	10.66
Carpet and Pad (Waste out of L/R)							-
Underlayment				54.93	sf	.75	41.20
Light				1		24.00	24.00
Smoke Detector				1		12.00	12.00
Door Bell							45.00
Thermostat							110.00
Door	2'6"	x	6'8"L	1		105.00	105.00
Door	2'6"	x	6'8"R	3		105.00	315.00
Door	2'	x	6'8"R	1		105.00	105.00
Door	2'	x	6'8"L	1		105.00	105.00
Paint Doors (2 sides)				6		30.00	180.00

Figure 7.5

Property Owner's Construction Company
123 School Street • Claremont, New Hampshire • (405) 732-1911
DETAILED REPAIR ESTIMATE

CUSTOMER	Allen Beck	INSURANCE COMPANY	LOSS CAUSE
ADDRESS	102 Aycock St.	ADDRESS	FILE #
	Newnan, GA	ADJUSTER	PHONE

DESCRIPTION	DIMENSIONS			QUANTITY	UNIT	UNIT COST	TOTAL COST
Bedroom: #1	10.33	12.33	8				
Sheetrock Walls and Ceiling				489.92	sf	.64	313.55
Paint Walls and Ceiling				489.92	sf	.24	117.58
Base Mold				45	lf	.97	43.65
Paint Base Mold				45	lf	.26	11.70
Carpet and Pad				16.66	yd	14.95	249.07
Double Window Unit	3'4"	x	4'2"	1		260.00	260.00
Paint Windows (2 sides)				2		30.00	60.00
Screens				2		26.00	52.00
Heat Vent				1		14.00	14.00
Closet:	2.5	6	8				
Sheetrock Walls and Ceiling				156	sf	.64	99.84
Paint Walls and Ceiling				156	sf	.24	37.44
Base Mold				17	lf	.97	16.49
Paint Base Mold				17	lf	.26	4.42
Carpet				3.66	yd	14.95	54.72
Shelves				24	lf	3.00	72.00
Bedroom: #2	11.6	13.87	8				
Sheetrock Walls and Ceiling				749.42	sf	.64	479.63
Paint Walls and Ceiling				749.42	sf	.24	179.86
Base Mold				73	lf	.97	70.81
Paint Base Mold				73	lf	.26	18.98
Carpet and Pad				18.66	yd	14.95	278.97
Window Unit	3'4"	x	4'2"	1			130.00
Paint Window (2 sides)				1		30.00	30.00
Heat Vents				2		14.00	28.00
Shelves				48	lf	3.00	144.00
Closet:	5	2.5	8				
Sheetrock Walls and Ceiling				132.50	sf	.64	84.80
Carpet							-
Base Mold				10	lf	.97	9.70
Paint Base Mold				10	lf	.26	2.60
Shelves				10	lf	3.00	30.00
Hall Closet:	3.17	2	8				
Clean, Seal and Paint Walls and Ceiling				89.06	sf	.52	46.31
Carpet							-
Paint Base Mold				10.34	lf	.26	2.69
Shelves				12	lf	3.00	36.00

Figure 7.5 (cont.)

Property Owner's Construction Company
123 School Street • Claremont, New Hampshire • (405) 732-1911
DETAILED REPAIR ESTIMATE

CUSTOMER	Allen Beck	INSURANCE COMPANY		LOSS CAUSE	
ADDRESS	102 Aycock St.	ADDRESS		FILE #	
	Newnan, GA	ADJUSTER		PHONE	

DESCRIPTION	DIMENSIONS			QUANTITY	UNIT	UNIT COST	TOTAL COST
Master Bedroom:	11.5	13.87	8				
Sheetrock Walls and Ceiling				564.42	sf	.64	361.23
Paint Walls and Ceiling				564.42	sf	.24	135.46
Base Mold				50.74	lf	.97	49.22
Paint Base Mold				50	lf	.26	13.00
Carpet and Pad				18.66	yd	14.95	278.97
Window Unit	3'4"	x	4'2"	1		130.00	130.00
Paint Windows				2		30.00	60.00
Heat Vents				2		14.00	28.00
Window	3'4"	x	3'2"	1		130.00	130.00
Closet:	2	5	8				
Sheetrock Walls and Ceiling				122	sf	.64	78.08
Paint Walls and Ceiling				122	sf	.24	29.28
Carpet							-
Shelves				26	lf	3.00	78.00
Base Mold				24	lf	.97	13.58
Paint Base Mold				14	lf	.26	3.64
Master Bath:	5.5	7	8				
Sheetrock Ceiling				38.50	sf	.64	24.64
Paint Ceiling				38.50	sf	.24	9.24
Heat Lamp				1		65.00	65.00
Wallboard 1/2 Wall				100	sf	1.25	125.00
Clean Tile Walls and Tub							175.00
Replace Fixtures on Tub							125.00
Replace Water Closet							110.00
Shower Rod							12.00
Soap Holder				1		14.00	14.00
Tile Board Tub Enclosure							125.00
Towel Rods				2		12.00	24.00
Medicine Cabinet							68.00
Sink and Fixtures							245.00
Shelves				4	lf	3.00	12.00
Window Unit	3'4"	x	3'2"				130.00
Paint Window (2 sides)							30.00
Door	2'	x	6'8"	R			105.00
Paint Door (2 sides)				1		30.00	30.00
4' Fluorescent Light							32.00

Figure 7.5 (cont.)

Property Owner's Construction Company
123 School Street • Claremont, New Hampshire • (405) 732-1911
DETAILED REPAIR ESTIMATE
FILE #

CUSTOMER	INSURANCE COMPANY	LOSS CAUSE
Allen Beck		
ADDRESS	ADDRESS	FILE #
102 Aycock St.		
Newnan, GA	ADJUSTER	PHONE

DESCRIPTION	DIMENSIONS			QUANTITY	UNIT	UNIT COST		TOTAL COST
Hall Bath:	5	7	8					
Sheetrock Ceiling				35	sf		.64	22.40
Paint Ceiling				35	sf		.24	8.40
Sheetrock 1/2 Wall				96	sf		.64	61.44
Paint 1/2 Wall				96	sf		.24	23.04
Tile 1/2 Wall and Tub Enclosure				98	sf		7.50	735.00
Tile Floor				35	sf		7.50	262.50
Tile Base				12	lf		4.50	54.00
Soap and Paper Holder				2			14.00	28.00
Water Closet								110.00
Sink and Fixtures								245.00
Bath Fan								32.00
Light				1			34.00	34.00
Clean Tub								18.00
Tub Fixtures								125.00
Laundry Room:	11.33	6.5	8					
Sheetrock Walls and Ceiling				381.42	sf		.64	244.11
Paint Walls and Ceiling				381.42	sf		.24	91.54
Base Mold				38	lf		.97	36.86
Paint Base Mold				38	lf		.26	9.88
Carpet				8.66	yd		14.95	129.47
Light				1			24.00	24.00
Water Heater Replaced								245.00
Door	2'6"	x	6'8"R	1			110.00	110.00
Paint Door (2 sides)				1			30.00	30.00
5/8" Plywood Underlayment				80.14	sf		.80	64.11
Shelves				14	lf		3.00	42.00
Fan								86.00
Closet:	5	2	8					
Sheetrock Walls and Ceiling				122	sf		.64	78.08
Paint Walls and Ceiling				122	sf		.24	29.28
Base Mold				14	lf		.97	13.58
Paint Base Mold				14	lf		.26	3.64
Vinyl				2	yd		14.95	29.90
Shelves				10	lf		3.00	30.00

Figure 7.5 (cont.)

Property Owner's Construction Company
123 School Street • Claremont, New Hampshire • (405) 732-1911

DETAILED REPAIR ESTIMATE

CUSTOMER	Allen Beck	INSURANCE COMPANY		LOSS CAUSE	
ADDRESS	102 Aycock St.	ADDRESS		FILE #	
	Newnan, GA	ADJUSTER		PHONE	

DESCRIPTION	DIMENSIONS			QUANTITY	UNIT	UNIT COST	TOTAL COST
Kitchen:	11.33	17x17	8				
Sheetrock Walls and Ceiling				650.53	sf	.64	416.34
Paint Walls and Ceiling				650.53	sf	.24	156.13
Base Mold				58	lf	.97	56.26
Paint Base Mold				58	lf	.26	15.08
Carpet and Pad				24	yd	14.95	358.80
Underlayment				194.53	sf	.80	155.62
Small Chandelier				1		85.00	85.00
Exterior Door	2'8"	x	6'8"L				225.00
Screen Door	2'8"	x	6'8"				89.00
Paint Doors				2		30.00	60.00
Double Window	3'4"	x	3'2"	1		260.00	260.00
Paint Window (2 sides)				2		30.00	60.00
Paint Screens				2		26.00	52.00
Base Cabinets				15	lf	75.00	1,125.00
Counter Tops				15	lf	20.00	300.00
Wall Cabinets				19.50	lf	75.00	1,462.50
Sink				1		165.00	165.00
Splash				19.50	lf	6.50	126.75
42" Vented Vent Hood							180.00
30" Stove							open
24" Fluorescent Light							24.00
Heat Vents				2		14.00	28.00
Clean Heating System							150.00
Ceiling Insulation				1232.25	sf	.42	517.55
Cornice Repair				24	lf	8.00	192.00
Siding Repair				2	sq	154.00	308.00
Decking Repair				64	sf	.85	54.40
Shingle Repair				2	sq	45.00	90.00
Paint Exterior Siding				1314	sf	.32	420.48
Paint Cornice				180	lf	2.00	360.00
Paint Screens				10		10.00	100.00
Pressure Wash Deck and Siding							240.00
Tear Out and Debris Removal							1,265.00
Permit							95.00
Temporary Power Cut onto House							75.00
Electrical Repair (Smith Electric)							715.00
						Subtotal	22,959.58
						20% O & P	4,591.92
						Total	$27,551.50

Figure 7.5 (cont.)

Chapter 8

COMMERCIAL CASE STUDY

This chapter is a case study of a commercial insurance repair construction project. It is a complete example, in the form of documentation and an explanation of the step-by-step procedures that the adjuster and insurance repair contractor(s) must carry out. An adjuster's report and estimate are provided, as are two separate contractor estimates, showing the different approaches and resulting cost differences for each. Correspondence between the parties is also shown, exactly as it would be written for an actual project. Contractors who familiarize themselves with the proper procedures, forms, and methods for preparing estimates are one step closer to profiting from insurance repair work.

The following paragraphs describe the series of events – from the original damage to the structure – to the investigations and arrangements made by the insurance company for our example project.

A single engine aircraft crashes into the east elevation of a building owned by R. Gray Enterprises in Atlanta, Georgia. A fire ensues and the third floor of the building is destroyed. The pilot is injured but the building is vacant at the time of the accident, so there are no other injuries.

The Federal Aviation Authority and the National Transportation Safety Board arrive three hours after the crash. No aircraft parts are found in the building, so the authorities release the premises the same day the crash occurs.

R. Gray's insurance company makes arrangements to have the building boarded up and secured by a local contracting firm, the T.R. Moore Company. The adjuster then inspects the site, and prepares his own estimate within a week of when the damage occurred.

The Moore Company surveys the site and then submits an estimate within ten days of the accident. Meanwhile, the insured procures their favored contractor, the Bradenton Construction Company, to prepare an estimate, also within ten days. As an incentive to choose their company, T.R. Moore offers to waive the $4,128 board-up fee, if they are awarded the job.

Next, the insurance company reviews the adjuster's report and estimate, and makes a decision for the amount of the claim within two to three weeks. In this case, the only limiting factor is the deductible, which is $50,000. The estimated price includes a loss of rents provision of $340,200. The total agreed price of the loss is estimated at $847,630.78. Since the adjuster does not know which contracting company will be used, the insurance company adds the $4,128 emergency board-up fee into the net claim payable, for a total of $851,758.78. The final payment agreed by the insurance company, applying the $50,000 deductible, is $801,758.78.

Ultimately, the T.R. Moore Company is awarded the work, because of its offer to waive the $4,128 board-up fee. The insurance company ends up paying $801,758.78, less $4,128, or $797,630.78.

The adjuster's report is shown in Figure 8.1. Figures 8.2 and 8.3 are the adjuster's takeoff and estimate. The contractors' priced quantity takeoffs are shown in Figures 8.4 and 8.5. The prices for these estimates were derived from Means *Building Construction Cost Data*, 1989. For a complete explanation of the estimating process, see Chapter 4 of this book, or refer to *Means Repair and Remodeling Estimating*.

Means Forms

CONDENSED
ESTIMATE SUMMARY

SHEET NO. *1*

PROJECT *R. Gray Enterprises*

ESTIMATE NO.

LOCATION *2001 J.C. Blvd., Norcross, Ca.* TOTAL AREA/VOLUME

DATE *4-10-89*

ARCHITECT

COST PER S.F./C.F.

NO. OF STORIES

PRICES BY: *As noted*

EXTENSIONS BY:

CHECKED BY: *Crosa*

	Bradenton	Adjuster	Moore
Demolition	28108	18105	18105
Concrete	40434	39934	39934
Metals	123240	124580	124580
Moisture – Thermal Control	22694	22648	22648
Finishes	306978	299484	299484
Equipment	1500	2130	2130
Plumbing	12000	12143	12143
HVAC	32000	31815	31815
Electrical	90600	90090	90090
Overhead	98633	64093	96140
Profit	151237	64093	110561
Totals	907424	769115	847630

(figures rounded off to nearest dollar)

Figure 8.1

OGLE & CROSA

Insurance Adjusters

P.O. BOX 16307
ATLANTA, GEORGIA 30321

(404) 969-7722

April 10, 1989

M. D. Dunavant
Claims Vice President
Merrimack Insurance Group
4447 Merrimack Plaza
Des Plaines, IL 60016

Reference: Claim No.:   4HB127063
                  Policy No.:   HPR 110079
                  Insured: R. Gray Enterprises
                  Date of Loss: 3-22-89
                  Our File No.:   PC-13601

Dear Ms. Dunavant:

This supplements our report of March 27, 1989. Since that time
we have developed enough information to provide you with this
report which includes settlement recommendations.

Enclosures

(1) Calculator Cost Form, Marshall Valuation Service
(2) Adjuster Takeoff
(3) T. R. Moore Co. Estimate ($847,630.78)
(4) Bradenton Construction Co. Estimate ($907,424.00)
(5) Adjuster's Estimate ($769,116.78)
(6) Statement of Loss

Reserves

It appears that your exposure under the building portion of this
claim will approach $800,000.00. Reserve accordingly.

With regard to Loss of Rents it is anticipated that the
restoration period will be approximately three months.

It is expected that rental revenue will not return to normal
until approximately three months after restoration. The affected
portions of the building yield monthly revenues of $56,700.
Therefore, a six month period of interruption would result in a
Loss of Rents claim of $340,200. Reserve accordingly.

EXECUTIVE OFFICES: 6611 WATSON STREET   •   UNION CITY, GEORGIA 30291

Figure 8.1 (cont.)

M. D. Dunavant
April 10, 1989
Page two

Coverage

Under Policy No. HPR 110079, The Merrimack Insurance Group
provided building coverage in the amount of $5.1 million dollars
on a three story building situated at 2001 Jimmy Carter Blvd. in
Norcross, Georgia 30091.

Applicable forms were HPR-101, 106 and RC-673. Coverage is
written on a replacement cost basis subject to a $50,000.00
deductible.

With regard to Loss of Rents your limit of liability is
$765,450.00.

Insured

The insured is R. Gray Enterprises, a Georgia corporation since
1967. The Secretary of State's office lists the following
principals: Mr. R. Gray, President; L. N. Gray, V.P.; W. R.
Tucker, Secretary/Treasurer. The primary source of revenue ($50
million in fiscal year 1988) is commercial real estate
development. Corporate headquarters are maintained in the
building which is the subject of this loss report.

Our contact has been with Mr. R. Gray. We found him to be
cooperative and conscientious in assisting us to develop his
loss. He was able to provide us with working blue prints of the
structure and has made all financial records available to us for
purposes of determining accurate Loss of Rents reserves.

Risk

The risk is a three story office building (80% occupied) situated
at 2001 Jimmy Carter Blvd. in Norcross, Georgia 30091. The
building was erected in 1982 and includes a basement parking
garage.

Construction consists of composite steel frame with an aluminum
panel and glass curtain wall system. Interior and exterior
premises are maintained in an excellent manner.

We prepared a Calculator Cost Form and arrived at a full
replacement cost value of $4,275,040.00. This is more than
10% less than the face value of your policy. Therefore, you
will want to refer the file to your underwriting department
for additional review.

Figure 8.1 (cont.)

M. D. Dunavant
April 10, 1989
Page three

Origin

The loss occurred at 0500 hours when a single engine aircraft crashed into the east elevation of the building at the highest northern-most corner. The aircraft careened off the building and into a nearby field.

At impact a fire ensued so that the eastern half of the third floor was engulfed in flames by the time fire crews arrived. The fire was controlled with the use of water cannons within 20 minutes of the fire department's arrival and totally extinguished by 05:42 hours.

Extensive damage was sustained as follows: third floor - direct damage by impact, fire, heat, smoke, soot and water; second floor -heat, smoke, soot, and water; first floor - smoke, soot and water.

Other than the pilot, no personal injury resulted. There was extensive damage to furnishings at the third and second floors. However, that property (owned by tenants and leasing companies) was covered under separate policies by a different insurer. You have no exposure with regard to contents.

Adjustment

As you are aware, we received this assignment via our emergency phone line by 07:15 hours on the day of the loss. We attempted immediate phone contact with the insured to no avail. By 08:11 hours we arrived at the scene and met him there.

Authorities had cordoned off the area and prevented any inspection of the premises pending arrival of federal agencies, the Federal Aviation Authority (FAA), and the National Transportation Safety Board (NTSB).

Meanwhile arrangements were made with the T. R. Moore Construction Co. to prepare for emergency board-up or lock-down once the green light was given by the federal authorities.

Besides fuel residue, no major aircraft components were found in the building so the federal authorities released the premises by 15:30 hours. Weather was sunny and dry and so this delay was not a problem.

Figure 8.1 (cont.)

M. D. Dunavant
April 10, 1989
Page four

Hazardous debris was removed and a tarpaulin was rigged over the upper story on the east end of the building. This was completed by 22:00 hours and served to protect the interior of the premises from the elements.

We began our survey immediately upon release of the building by the federal authorities. The results of this preliminary survey as well as 48 photos were sent to you in our report of March 27, 1989.

Since that time, as you instructed, we retained the services of a professional engineering firm to determine to what extent, if any, the structural soundness of the building had been impacted. Their report was sent to you under separate cover. The enclosed estimate was written so as to incorporate the findings of the P.E. report.

Our takeoff is enclosed and is the basis for our final estimate. This came to $769,116.78. The insured had called in a contractor of his choice (Bradenton Construction Co.) for an additional opinion. We called T. R. Moore Construction Co. for a third estimate.

The adjuster's takeoff was provided to both construction firms for price continuity. A few items were added to the adjuster's takeoff after all three estimators concurred that there had been some inadvertent omissions.

The T. R. Moore Construction Co. provided an estimate in the amount of $847,630.78 which did not include the costs for emergency "board up" performed shortly after the loss occurred. This came to $4,128. Moore is willing to waive this charge if his firm is awarded the restoration job.

The Bradenton Construction Co. submitted an estimate in the amount of $907,424.00. Although the insured expressed a preference for this firm, he made it clear that he would work with either contractor provided a surety bond was obtained. He understood that settlement would be based on the most reasonable figure.

We have prepared a condensed estimate summary showing a comparison of the three enclosed estimates by category of operation. You will see that the various categories compared favorably. A major difference was found in the area of overhead and profit. Our estimate figured the standard for the insurance industry which is 20 percent. However, this project involves

Figure 8.1 (cont.)

M. D. Dunavant
April 10, 1989
Page five

extenuating circumstances such as the fact that much of the restoration is at the third story and includes extensive exterior work. This will require additional equipment and risks. Also there is the expense of surety bonding not usually required in smaller insurance repair losses. Therefore we feel that a departure from the standard allowance for O&P is in order.

The Bradenton Construction Co. estimate is calling for 15% overhead and 20% profit. These charges alone come to $249,870.00.

T. R. Moore's estimate calls for 15% overhead but then he is willing to reduce profit to 15% for a total O&P charge of $206,700.13. Your company stands to gain an additional savings of $4,128.00 due to Moore waiving the emergency restoration costs if he is awarded this project.

As yet, no decision has been made as to which contractor will perform the repairs. On the possibility that Bradenton Construction Co. will agree to enter into a contract for the lower repair figure, you should prepare to issue the draft for emergency services rendered in the amount of $4,128.00 payable to T. R. Moore Construction Co. We will let you know within the next seven days if this payment will be called for.

The attached Statement of Loss summarizes details as to values at risk, actual loss, and claim payable. The actual loss is based on the total of the figure submitted by T. R. Moore coupled with the emergency restoration cost of $4,128.00 for a total of $851,758.78. The only limiting factor is the $50,000 deductible which reduces the net claim payable to $801,758.78

Therefore, we are recommending that you extend authorization to us to secure a "Proof of Loss" from the insured in the amount of $801,758.78. We have discussed this figure with the insured and he is prepared to settle on that basis.

Subrogation and Salvage

Pending further investigation by the FAA and NTSB, subrogation possibilities look good. We are told that a final determination as to the cause, be it pilot error or mechanical failure, will be forthcoming in the next 45 days. The aircraft liability carriers have been placed on notice of your company's interest in indemnification.

Figure 8.1 (cont.)

M. D. Dunavant
April 10, 1989
Page six

With regard to salvage, what was salvageable will be used in restoration. There may be some junk value in structural steel to be removed from the site, but this is not expected to net over $1,000.00. We will look into this and advise.

Title and Encumbrances

The property is mortgaged by The Fidelity Fiduciary Bank. Balance on the loan is $2.2 million. A check at the Gwinnett County Courthouse shows no liens, pending lawsuits, nor any other encumbrance.

Recommendation and Remarks

Again, we hereby request your authorization to secure a "Proof of Loss" as noted above. Upon receipt of the "Proof of Loss" properly signed and notarized by the insured, we will forward same to you with a request for draft issuance.

We will continue with our handling as indicated and keep you informed of all progress.

Very truly yours,

Peter J. Crosa, AIC
Ogle & Crosa

PJC:mdc

Enclosures

Figure 8.1 (cont.)

COST ANALYSIS

SHEET NO. **1**

PROJECT **R. Gray Enterprises, Inc.**

ESTIMATE NO.

ARCHITECT

DATE **3-22-89**

TAKE OFF BY: **Crosa** QUANTITIES BY: PRICES BY: EXTENSIONS BY: CHECKED BY:

DESCRIPTION	SOURCE/DIMENSIONS			QUANTITY	UNIT	MATERIAL		LABOR		EQ./TOTAL	
						UNIT COST	TOTAL	UNIT COST	TOTAL	UNIT COST	TOTAL
Tear out and debris removal as follows:											
① −roof over layments −concrete plank roof deck −steel beams −precast columns − shore up-deck				14,700	CF						
② − haul away				1244	CY						
③ − partition walls at 3rd & 2nd flr.				27,825	CF						
Concrete											
④ replace roof planks (4" × 13")	60'	×	70'	4200	SF						
⑤ concrete column 16" round 12' ht.	3	@	12'	2	CY						
⑥ replace prestressed floor slab 4" deep	60'	×	70'	4200	SF						
Metals											
⑦ replace steel beams 14"×30"×35'	2	@	35	70	LF						
⑧ replace curtain wall- alum./glass	12'	×	130	1560	SF						
Roofing											
⑨ replace fiberglass insulation 2¼ R8	70'	×	90'	6300	SF						
⑩ install cold applied roofing 3 ply sys. with ceramic grain coating	105	×	90	94.5	SQ						
⑪ cost differential for 3rd floor (40%)											

Figure 8.2

**COST
ANALYSIS**

SHEET NO. **3**

PROJECT **R. Gray Enterprises**

ESTIMATE NO.

ARCHITECT

DATE **3-22-89**

TAKE OFF BY: *Crosa*   QUANTITIES BY: PRICES BY: EXTENSIONS BY: CHECKED BY:

DESCRIPTION	SOURCE/DIMENSIONS			QUANTITY	UNIT	MATERIAL		LABOR		EQ./TOTAL	
						UNIT COST	TOTAL	UNIT COST	TOTAL	UNIT COST	TOTAL
HVAC											
㉑ Replace HVAC complete at 1/3 of 3rd floor including duct system, registers and overhaul air handler (rule of thumb)				6300	SF						
Electrical											
㉒ Replace complete circuits, light fixtures, outlets, & check service panel as follows:											
– 3rd floor 1/3				6300	SF						
– 2nd floor 1/3				6300	SF						
– 1st floor 1/4				4725	SF						
(rule of thumb)											

Figure 8.2 (cont.)

133

SHEET NO. **2**

PROJECT **R. Gray Enterprises**

ESTIMATE NO.

ARCHITECT

DATE **3-22-89**

TAKE OFF BY: **Crosa**   QUANTITIES BY: PRICES BY: EXTENSIONS BY: CHECKED BY:

DESCRIPTION	SOURCE/DIMENSIONS			QUANTITY	UNIT	MATERIAL		LABOR		EQ./TOTAL	
						UNIT COST	TOTAL	UNIT COST	TOTAL	UNIT COST	TOTAL
Interior											
⑫ replace complete carpeting - 3 floors incl. 70 oz. pad & installation	210	x	90	6480	SY						
⑬ replace partition walls as follows: −5/8" drywall - taped both sides on metal studs 16"O.C. 2nd & 3rd floors	1,010'	x	12'	12,120	SF						
⑭ clean & paint remaining walls at 2nd & 3rd floor	1,010	x	12	12,120	SF						
⑮ clean & paint walls at first floor	1,010	x	12	12,120	SF						
⑯ replace suspended fiberglass ceiling (panels 2'x4'x 3/4") incl. suspension sys. all 3 floors	3	x	18900	56,700	SF						
Equipment											
⑰ crane incl. oper.	4	@	270	4	day						
⑱ forklift incl. oper.	3	@	210	3	day						
⑲ scaffolding				140	sec.						
Plumbing											
⑳ reinstall all systems at eastern quarter of 3rd floor (rule of thumb)				4725	SF						

Figure 8.2 (cont.)

Means Forms

COST ANALYSIS

SHEET NO. **1**

PROJECT **R. Gray Enterprises, Inc.**

ESTIMATE NO.

ARCHITECT

DATE **3-22-89**

TAKE OFF BY: **Crosa**   QUANTITIES BY: PRICES BY: EXTENSIONS BY: CHECKED BY:

DESCRIPTION	SOURCE/DIMENSIONS			QUANTITY	UNIT	MATERIAL		LABOR		EQ./TOTAL	
						UNIT COST	TOTAL	UNIT COST	TOTAL	UNIT COST	TOTAL
Tear out and debris removal as follows:											
①-roof over layments											
-concrete plank roof deck											
- steel beams				14,700	CF					.24	3528.
- precast columns											
- shore up - deck											
②- haul away				1244	CY					6.35	7899.40
③- partition walls at 3rd ¢ 2nd flr.				27,825	CF					.24	6678.
Concrete											
④replace roof planks (4"x 13")	60'	x	70'	4200	SF					5.30	22260.
⑤concrete column 16" round 12' ht.	3	@	12'	2	CY IM						1000.
⑥replace prestressed floor slab 4" deep	60'	x	70'	4200	SF					3.97	16674.
Metals											
⑦replace steel beams 14"x 30"x35'	2	@	35	70	LF					19.15	1340.50
⑧replace curtain wall-alum./glass	12'	x	130	1560	SF					79.	123240.
Roofing											
⑨replace fiberglas insulation 2¼ R8	70'	x	90'	6300	SF					1.18	7434.
⑩install cold applied roofing 3 ply sys. with ceramic grain coating	105	x	90	94.5	SQ					115.	10867.50
⑪cost differential for 3rd floor (40%)											4347.
Page Total											205268.40

Figure 8.3

135

COST ANALYSIS

SHEET NO. **2**

PROJECT **R. Gray Enterprises, Inc.**

ESTIMATE NO.

ARCHITECT

DATE **3-22-89**

TAKE OFF BY: **Crosa** QUANTITIES BY: PRICES BY: EXTENSIONS BY: CHECKED BY:

DESCRIPTION	SOURCE/DIMENSIONS			QUANTITY	UNIT	MATERIAL UNIT COST	TOTAL	LABOR UNIT COST	TOTAL	EQ./TOTAL UNIT COST	TOTAL
Interior											
⑫ replace complete carpeting- 3 floors incl. 70 oz. pad & installation	210	×	90	6480	SY					26.5	171720.
⑬ replace partition walls as follows: - 5/8" drywall taped both sides on metal studs 16"o.c. 2nd & 3rd floors	1,010	×	12'	12,120	SF					2.51	30421.20
⑭ clean & paint remaining walls at 2nd & 3rd floor	1,010	×	12	12,120	SF					.32	3878.40
⑮ clean & paint walls at first floor	1,010	×	12	12,120	SF					.32	3878.40
⑯ replace suspended fiberglas ceiling (panels 2'x4'x 3/4") incl. suspension sys. all 3 floors	3	×	18,900	56,700	SF					1.58	89586.
Equipment											
⑰ crane incl. oper.	4	@	270	4	day					270	1080.
⑱ forklift incl. oper.	3	@	210	3	day					210	630.
⑲ scaffolding				140	sec					3.	420.
Plumbing											
⑳ reinstall all systems at eastern quarter of 3rd floor (rule of thumb)				4725	SF					2.57	12143.25
Page Total											313757.25

Figure 8.3 (cont.)

SHEET NO. **3**

PROJECT **R. Gray Enterprises, Inc.**

ESTIMATE NO.

ARCHITECT

DATE **3·22·89**

TAKE OFF BY: **Crosa**   QUANTITIES BY: PRICES BY: EXTENSIONS BY: CHECKED BY:

DESCRIPTION	SOURCE/DIMENSIONS			QUANTITY	UNIT	MATERIAL		LABOR		EQ./TOTAL	
						UNIT COST	TOTAL	UNIT COST	TOTAL	UNIT COST	TOTAL
HVAC											
21) replace HVAC complete at 1/3 of 3rd floor incl. duct system, registers and overhaul air handler											
(rule of thumb)				6300	SF					5.05	31815.
Electrical											
22) replace complete circuits, light fixtures, outlets, & check service panel as follows:											
— 3rd floor 1/3				6300	SF					5.20	32760.
— 2nd floor 1/3				6300	SF					5.20	32760.
— 1st floor 1/4				4725	SF					5.20	24570.
(rule of thumb)							Page Total				121905.
							All Pages Total				640930.65
							Contractors Overhead and Profit (20%)				128186.13
							Total				769116.78

Figure 8.3 (cont.)

T. R. MOORE COMPANY
GENERAL CONTRACTORS
710 51 St. N.E.
Atlanta, Georgia 30303

April 3, 1989

Peter J. Crosa
Ogle & Crosa
PO Box 16307
Atlanta, GA 30321

Reference: R. Gray Enterprises, INc.
                2001 Jimmy Carter Blvd.
                Norcross, Georgia

Dear Mr. Crosa:

In accordance with the takeoff you have provided, we are supplying
you with our priced estimate to restore the fire loss at the above
premises.

Prices include labor and materials. We are surety bonded and carry
workers compensation insurance.

Please be advised that, if awarded this restoration project, we
will waive the cost of $4,128.00 incurred for emergency services
on the date of the fire.

TEAR OUT AND DEBRIS REMOVAL AS FOLLOWS:
1. Roof over layments
    Concrete plank roof deck
    Steel beams
    Pre-cast columns
    Shore up - deck   14,700 cf   @ $0.24 $3,528.00
2. Haul away   1,244 cf   @ 6.35   7,899.40
3. Partition walls at
    3rd and 2nd floors   27,825 cf   @ .24   6,678.00

CONCRETE:
4. Replace roof planks
    (4" x 13")   4,200 sf   @ 5.30   22,260.00
5. Concrete column
    16" round, 12' ht.   2 cy   1,000.00
6. Replace pre-stressed
    floor slab 4" deep   4,200 sf   @ 3.97   16,674.00

METALS:
7. Replace steel beams
    14" x 30" x 35'   70 lf   @ 19.15   1,340.50
8. Replace curtain wall
    aluminum/glass   1,560 sf   @ 79.00 123,240.00

Figure 8.4

138

Mr. Peter J. Crosa
April 3, 1989
Page two

ROOFING:
9. Replace fiberglass insulation 2 1/4 R8	6,300 sf	@ 1.18	7,434.00
10. Install cold applied roofing 3-ply system with ceramic grain coating	94.5 sq	@ 115.00	10,867.50
11. Cost differential for 3rd floor (40%)			4,347.00

INTERIOR:
12. Replace complete carpeting 3 floors incl. 70 oz. pad and installation	6,480 sy	@ 26.50	171,720.00
13. Replace partition walls as follows: 5/8" drywall - taped both sides on metal studs 16" o.c.	12,120 sf	@ 2.51	30,421.20
14. Clean and paint remaining walls at 2nd and 3rd floor	12,120 sf	@ .32	3,878.40
15. Clean and paint walls at 1st floor	12,120 sf	@ .32	3,878.40
16. Replace suspended fiber-glass ceiling (panels 2'x4'x3/4") including suspension system - all 3 floors	56,700 sf	@ 1.58	89,586.00

EQUIPMENT:
17. Crane including operator	4 day	@ 270.00	1,080.00
18. Forklift including operator	3 day	@ 210.00	630.00
19. Scaffolding	140 sec	@ 3.00	420.00

PLUMBING:
20. Reinstall all systems at eastern quarter of 3rd floor (rule of thumb)	4,725 sf	@ 2.57	12,143.25

HVAC:
21. Replace HVAC complete at 1/3 of 3rd floor including duct system, registers and overhaul air handler (rule of thumb)	6,300 sf	@ 5.05	31,815.00

Figure 8.4 (cont.)

Mr. Peter J. Crosa
April 3, 1989
Page three

ELECTRICAL:
22. Replace complete circuits,
    light fixtures, outlets &
    check service panel as
    follows:

3rd floor	1/3	6,300 sf	@ 5.20	32,760.00
2nd floor	1/3	6,300 sf	@ 5.20	32,760.00
1st floor	1/4	4,725 sf	@ 5.20	24,570.00
(Rule of thumb)				

SUBTOTAL:	$640,930.65
OVERHEAD (15%):	96,139.60
SUBTOTAL:	$737,070.25
PROFIT (15%):	110,560.54
GRAND TOTAL:	$847,630.79

Sincerely yours,

T. R. Moore

TRM:mdc

Figure 8.4 (cont.)

Bradenton Construction Company
1220 McCartney Blvd.
Bradenton, Georgia 30136

April 3, 1989

Mr. R. Gray, President
R. Gray Enterprises
Norcross, Georgia

Dear Mr. Gray:

It is our pleasure to submit this bid to restore your building as per the specifications provided by your insurance company.

1.	Demolition including built up roofing, concrete decks, I-beams, and pre-cast concrete columns and floors. Also include shoring of third floor.	$28,108
2.	Concrete work including pre-cast roof deck, columns and floor slab	40,434
3.	Replace aluminum curtain wall incl. glazing	123,240
4.	Replace insulation on roof	7,434
5.	Replace built up roof with gravel top	10,900
6.	Height differential at 3rd story	4,360
7.	Replace carpeting including pad and installation - 2nd and 3rd floor 6,720 sq. ft. @ $26.50	178,080
8.	Replace partition walls; drywall on metal - collapsible 12,120 sq. ft. @ $2.55	30,906
9.	Clean and paint remaining drywall at all three floors	7,272
10.	Replace suspended ceiling 2x4 acoustical - all floors 56,700 sq. ft. @ $1.60	90,720
11.	Heavy equipment rental including crane and towmotor	1,500

Figure 8.5

141

Mr. R. Gray
April 3, 1989
Page two

12. Plumbing   12,000

13. HVAC including duct work   32,000

14. Electrical - complete including
    strip lighting - all floors   90,600
                                                    $657,554
                                   Overhead 15%   98,633
                                   Subtotal $756,187
                                   Profit 20%   151,237
                                   TOTAL: $907,424

Respectfully submitted,

J. Lennon

J. Lennon, President

JL:mdc

Figure 8.5 (cont.)

APPENDIX

National Associations

National Association of
Independent Insurance
Adjusters
222 W. Adams Street
Chicago, IL 60606

National Association of
Marine Surveyors
86 Windsor Gate Drive
North Hill, NY 11040

National Association of
Insurance Women
P.O. Box 4410
Tulsa, OK 74159

National Association of
Catastrophe Adjusters, Inc.
P.O. Box 960
Comfort, TX 78013

Truck and Heavy Equipment
Claims Council
10054 Willdan
St. Louis, MO 63123

Alabama

Alabama Claims Association
P.O. Box 6190 C
Birmingham AL 35209

Mobile Claims Association
P.O. Box 9546
Mobile, AL 36691

Montgomery Claims
Association
P.O. Box 17399
Montgomery, AL 36117

Arizona

Arizona Insurance Claims
Association
P.O. Box 11331
Phoenix, AZ 85061

Southern Arizona Claims
Association
P.O. Box 12663
Tucson, AZ 85732

Phoenix Section of
Arizona Insurance Claims
Association
P.O. Box 11331
Phoenix, AZ 85031

Arkansas

Arkansas Adjuster's
Association
P.O. Box 978
No. Little Rock, AR 72110

Fort Smith Claims Association
P.O. Box 626
Fort Smith, AR 72902-0626

California

California Association of
Independent Insurance
Adjusters
P.O. Box 297
Salinas, CA 93902

Central California Adjuster's
Association
2974 N. Fresno Street
Suite 542
Fresno, CA 93726

Central Coast Claims
Association
P.O. Box 6388
San Jose, CA 95150

Mid-Valley Claims Association
1209 N. Center St.
Stockton, CA 95202

North Coast Adjuster's
Association
P.O. Box 1299
Healdsburg, CA 95448

Orange County Adjuster's
Association
P.O. Box 11204
Santa Ana, CA 92711

San Fernando Valley
Adjuster's Association
P.O. Box 242
Canoga Park, CA 91303

San Gabriel Valley Claims
Association
P.O. Box 3337
Chatsworth, CA 91313-3337

Workers Compensation
Claims Association
P.O. Box 10501
Burbank, CA 91510-0501

Riverside-San Bernadino
Counties Adjuster's
Association
370 W. 6th Street, Suite 115
San Bernadino, CA 92401

San Diego Insurance
Adjuster's Association
2550 5th Ave., Suite 826
San Diego, CA 92103

Casualty Claims Association of
San Francisco
605 3rd St. , Suite 203
San Francisco, CA 94120

Surety Claims Association of
San Francisco
450 Sansome Street
San Francisco, CA 94111

Colorado

Colorado Claims Association
280 W. 44th Ave.
Denver, CO 80211

Colorado Springs Claims
Association
P.O. Box 7173
Colorado Springs, CO 80933

Denver Claims Association
2701 Alcott St., Room 471
Denver, CO 80211

Connecticut

Connecticut Association of
Independent Insurance
Adjusters
P.O. Box 72
New Haven, CT 06473

Delaware

Delaware Claims Association
P.O. Box 5070
Wilmington, DE 19808

District of Columbia

Washington Claims
Association
P.O. Box 183
Springfield, VA 22152

Florida

Brevard County Claims
Association
P.O. Box 7
Cocoa, FL 32922

Pinellas County Claims
Association
P.O. Box 20367
St. Petersburg, FL 33742

Sarasota-Bradenton Claims
Association
P.O. Box 4175
Sarasota, FL 33578

South Florida Claims
Association
P.O. Box 650637
Miami, FL 33165

Ft. Lauderdale Claims
Association
2430 W. Oakland Park Blvd.
Ft. Lauderdale, FL 33311

Orlando Claims Association
P.O. Box 20771
Orlando, FL 32814

Georgia

Atlanta Claims Manager's
Council
4 Piedmont Center
3565 Piedmont Road
Atlanta, GA 30305

Columbus Claims Association
P.O. Box 4239
Columbus, GA 31904

Macon Claims Association
P.O. Box 7652
Macon, GA 31204

Hawaii

Hawaii Claims Association
P.O. Box 1140
Honolulu, HI 96807

Hawaii Claims Managers
Association
1833 Kalakaua Ave., Suite 602
Honolulu, HI 96815

Big Island Claims Association
P.O. Box 1899
Hilo, HI 96720

Idaho

Boise Association of Idaho
Adjusters
P.O. Box 199
Boise, ID 83701

Illinois

Illinois Association of
Independent Insurance
Adjusters
1207 Sunset Road
Wheaton, IL 60187

Northeastern Illinois
Adjuster's Association
230 W. Monroe St.
Chicago, IL 60606

Quad-City Claims Association
P.O. Box 960
Bettendorf, Iowa 52722

South Illinois Adjuster's
Association
P.O. Box 463
Marion, IL 62959

Southwestern Illinois Claims
Association
407 E. Lincoln
Belleville, IL 62220

Casualty Adjuster's
Association of Chicago
212 W. Washington, HQ 2H
Chicago, IL 60606

Greater Peoria Claims
Association
3100 N. Knoxville
Peoria, IL 61604

Indiana

Indiana Association of
Independent Insurance
Adjusters
P.O. Box 2068
Evansville IN 47714

State Adjuster's Association of
Indiana, Inc.
P.O. Box 1883
Indianapolis, IN 46206

Indiana Association of Fire
Adjusters
1752 N. Meridian Street
Indianapolis, IN 46206

No. Central Adjusters
Association of Indiana
P.O. Box 263
Walton, IN 46994

So. Central Adjusters
Association of Indiana
P.O. Box 1149
Columbus, IN 47202

Tri-State Association of
Insurance Adjusters
P.O. Box 5524
Evansville, IN 47715

Fort Wayne Insurance
Adjuster's Association
P.O. Box 5412
Fort Wayne, IN 46895

Lafayette Adjuster's
Association
P.O. Box 55507
Lafayette, IN 47903

South Bend Claims
Association
P.O. Box 504
Mishawaka, IN 46544

Iowa

Iowa Casualty & Surety
Claims Association
P.O. Box 1336
Des Moines, IA 50305

Northeast Iowa Adjuster's
Association
P.O. Box 390
Waterloo, IA 50704

Quad-City Claims Association
P.O. Box 828
Bettendorf, IA 52722

Cedar Rapids Claims
Association
3480 10th Ave.
Marion, IA 52302

Kansas

Kansas Claims Association
P.O. Box 11281
Wichita, KS 67202

Mid-Kansas Claims
Association
P.O. Box 1401
McPherson, KS 67460

Topeka Claims Association
1000 W. 10th Street
Topeka, KS 66604

Wichita Claims Association
P.O. Box 3222
Wichita, KS 67201

Kentucky

Bluegrass Claims Association
P.O. Box 7259
Lexington, KY 40523

Bluegrass Association of
Insurance Women
5448 Bryan Station Road
Paris, KY 40361

Huntington Tri-State Claims
Association
P.O. Box 246
Huntington, W.Virginia 25701

Kentucky Claims Association
P.O. Box 8132
Lexington, KY 40533

West Kentucky Adjuster's
Association
P.O. Box 1795
Owensboro, KY 42302-1795

Greater Louisville Claims
Association
P.O. Box 893
Louisville, KY 40201

Louisiana

Independent Adjusters of
Louisiana
P.O. Box 9806
New Iberia, LA 70560

Northwest Louisiana Claims
Association
P.O. Box 990
Shreveport, LA 71104

Northeast Louisiana Claims
Association
P.O. Box 2444
Monroe, LA 71201

Southwest Louisiana Claims
Association
P.O. Box 33084
Lafayette, LA 70503

Baton Rouge Claims
Association
P.O. Box 14889
Baton Rouge, LA 70898

New Orleans Claims
Association
P.O. Box 791053
New Orleans, LA 70109-1053

Maine

Northern Maine Adjusters
Association
P.O. Box 1206
Bangor, ME 04401

Maryland

Baltimore Claimsmen's
Association
P.O. Box 9772
Baltimore, MD 21204

Massachusetts

Massachusetts Association of
Independent Insurance
Adjusters
P.O. Box 446
Brockton, MA 02403

New England Claims
Association
P.O. Box 300
Stamford, Connecticut 06904

Springfield Claims Manager's
Council
1111 Elm Street
W. Springfield, MA 01089

Boston Association of Claims
Executives
181 Wells Avenue
Newton, MA 02159

Michigan
Michigan Adjusters
Association
P.O. Box 8209
Grand Rapids, MI 49508

Michigan Association of
Independent Adjusters
P.O. Box 2284
Southfield, MI 48037

Minnesota
Northwest Loss Association
4640 W. 77th Street
Minneapolis, MN 55435

Twin Cities Claims Manager's
Council
8440 Normandale Lake Blvd.
Suite 500
Bloomington, MN 55437

Twin Cities Claims
Association
4600 W. 77th Street., Suite 197
Edina, MN 55435

Mississippi
Mississippi Claims Association
P.O. Box 55400
Jackson, MS 39216

Delta Chapter
Mississippi Claims Association
P.O. Box 1815
Greenville, MS 38702

Northeast Mississippi Claims
Association
P.O. Box 590
Tupelo, MS 38801

Missouri
Tri-State Claims Association
P.O. Box 2216, Station A
Joplin, MO 64803

Kansas City Claims
Association
P.O. Box 559
Kansas City, MO 64141

Springfield Claims Association
P.O. Box 4787
Springfield, MO 65808

St. Louis Claim Manager's
Council
No. One Mercantile Center
St. Louis, MO 63101

Montana
Montana State Adjuster's
Association
P.O. Box 3465
Bozeman, MT 59772

Billings Adjuster's Association
P.O. Box 20699
Billings, MT 59104

Nebraska
Nebraska Claims Association
P.O. Box 31125
Omaha, NE 68131

Mid-Nebraska Claims
Association
P.O. Box 986
Grand Island, NE 68802

Lincoln Claims Association
P.O. Box 83052
Lincoln, NE 68501

Nevada
Nevada State Claims
Association
5270 Neil Road, Suite 100
Reno, NV 89502

Northern Nevada Claims
Association
P.O. Box 71416
Reno, NV 89570-1416

Southern Nevada Claims
Association
P.O. Box 7529
Las Vegas, NV 89101

Reno Claims Association
P.O. Box 71316
Reno, NV 89570-1316

New Hampshire
New Hampshire Adjuster's
Association
543 West Street
Keene, NH 03431

Seacoast Adjuster's
Association of New
Hampshire
147 Middle Street
Portsmouth, NH 03801

New Jersey
Independent Insurance
Adjuster's Association
of New Jersey
100 Summit Avenue
Summit, NJ 07901

Delaware Valley Claims
Association
P.O. Box 6486
Lawrenceville, NJ 08648

Atlantic City Claims
Association
25 N. Hanover Ave.
Margate, NJ 08402

New Mexico
New Mexico Claims
Association
P.O. Box 20454
Albuquerque, NM 87154-0454

New York
New York Association of
Independent Adjusters
P.O. Box 1920
Ocean Side, NY 10007

New York Claims Association
119 Church Street
New York, NY 10007

Property Loss Association of
New York
116 John Street
New York, NY 10038

Mid-Hudson Claims
Association
Professional Building – Rt. 32
Central Valley, NY 10903

Albany Claims Association
P.O. Box 3615
Kingston, NY 14202

Buffalo Claim Manager's
Council
60 Lakefront Blvd.
Buffalo, NY 14202

Insurance Club of Buffalo
197 Delaware Ave.
Buffalo, NY 14202

Glens Falls Claims Association
333 Glens Street
Glens Falls, NY 12801

Oneonta Claims Association
P.O. Box 50
Oneonta, NY 13820

Rochester Claims Association
One Marine Midland Plaza
Rochester, NY 14604

Rochester Claims Manager's
Council
P.O. Box 4855
Syracuse, NY 13221

North Carolina
North Carolina Adjuster's
Association
978 Greenville Blvd.
Greenville, NC 27858

Neuse Adjuster's Association
P.O. Box 1712
Goldsboro, NC 27533

Northwest Claims Association
P.O. Box 2052
Hickory, NC 28601

Asheville Claims Association
P.O. Box 2712
Asheville, NC 28807

Charlotte Claims Association
P.O. Box 37194
Charlotte, NC 28237-7194

Fayetteville Claims Association
Drawer 2359
Fayetteville, NC 28302

Greensboro Claims
Association
P.O. Box 13228
Greensboro, NC 27420

Greenville Claims Association
P.O. Box 7186
Greenville, NC 27834

Raleigh Claims Association
P.O. Box 27427
Raleigh, NC 27611

Rocky Mount Claims
Association
P.O. Drawer 8105
Rocky Mount, NC 27804-8105

Wilmington Claims
Association
P.O. Box 5374
Wilmington, NC 28403

Winston-Salem Claims
Association
P.O. Box 5660
Winston-Salem, NC 27113

Ohio

Ohio Association of
Independent Insurance
Adjusters
3340 Hader Ave.
Cincinnatti, OH 45211

Ohio State Claims Association
P.O. Box 581
Columbus, OH 43216

Lorain County Claims
Association
P.O. Box 1091
Elyria, OH 44036

North Central Ohio
Association of Claims Men
Mansfield, OH 44900

Ohio Valley Claims
Association
P.O. Box 386
St. Clairsville, OH 43950

Southern Ohio Claims
Association
P.O. Box 933
Gallipolis, OH 45631

Akron Association of
Claims Men
P.O. Box 7155
Akron, OH 44306

Canton Claims Association
P.O. Box 8473
Canton, OH 44711

Cincinnati Claims Association
11306 Southland Road
Cincinnati, OH 45240

Cleveland Claims Association
4194 Rocky River Drive
Cleveland, OH 44135

Dayton Claims Adjuster's
Association
P.O. Box 334
Dayton, OH 45401

Toledo Claims Association
4841 Monroe
Toledo, OH 43606

Oklahoma

Oklahoma Claims Association
P.O. Box 6334
Moore, OK 73153

Ada Claims Men's Association
P.O. Box 1311
Ada, OK 74820

Lawton Claims Association
P.O. Box 1677
Lawton, OK 73502

Oklahoma City Claims
Association
P.O. Box 18817
Oklahoma City, OK 73154

Tulsa Claims Association
P.O. Box 54242
Tulsa, OK 74155-0242

Oregon

Oregon Casualty Adjuster's
Association
P.O. Box 2149
Lake Oswego, OR 97034

Independent Insurance
Adjusters of Oregon
10151 SW Barbur Blvd.
Suite 107D
Portland, OR 97223

Rogue Valley Insurance
Adjuster's Association
P.O. Box 1431
Medford, OR 97501

Pennsylvania

Pennsylvania Claims
Association
100 Laurel Lane, RD 3
Dalton, PA 18414

Pennsylvania Association of
Independent Insurance
Adjusters
207 Long Lane
Upper Darby, PA 19082

Bucktail Claims Association
P.O. Box 912
Warren, PA 16365

Lehigh Valley Claims
Association
P.O. Box 886
Allentown, PA 18105

Lower Bucks–Philadelphia
Claims Association
339 Market Street
Philadelphia, PA 19106

Midwestern Claims
Association
P.O. Box 2189
Butler, PA 16003

Susquehanna Valley Claims
Association
109 Oak Street
Danville, PA 17821

Tri-County Claims Association
P.O. Box 85
Windber, PA 15963

Washington County Claims
Association
775 E. Maiden Street
Washington, PA 15301

West Branch Claims
Association
P.O. Box 62
Montoursville, PA 17754

Erie Claims Association
P.O. Box 1699
Erie, PA 16530

Greensburg Claims
Association
37 McMurray Road
Pittsburgh, PA 15241

Harrisburg Claims Association
P.O. Box 2037
Harrisburg, PA 17105

Lancaster Valley Claims
Association
P.O. Box 127
E. Petersburg, PA 17520

Pittsburgh Claims Association
225 McAllister Drive
Pittsburgh, PA 15235

Reading Claims Association
P.O. Box 1219
Reading, PA 19603

Rhode Island

Independent Insurance
Adjusters of Rhode Island
111 Westminster
Providence, RI 02903

South Carolina

South Carolina Claims
Association
P.O. Box 14794
Surfside Beach, SC 29587

Claims Management
Association of South Carolina
P.O. Box 31190
Charleston, SC 29417

Property Damage Claims
Association
192 South Cashua Drive
Florence, SC 29502

Independent Insurance
Adjusters of South Carolina
P.O. Box 5712
Greenville, SC 29606

Piedmont Claims Association
P.O. Box 1264
Columbia, SC 29602

Charleston Area Claims
Association
P.O. Drawer 32159
Charleston, SC 29417

Columbia Claims Association
P.O. Box 1415
Columbia, SC 29202

South Dakota

South Dakota Claims
Association
1908 W. 42nd Street
Sioux Falls, SD 57105

Aberdeen Claims Association
P.O. Box 1326
Aberdeen, SD 57401

Tennessee

Tennessee Claims Association
P.O. Box 78
Elizabethton, TN 37643

Tri-City Claims Association
412 Village Drive
Greenville, TN 37743

Chattanooga Claims
Association
P.O. Box 8687
Chattanooga, TN 37411

Cookville Claims Association
P.O Box 2647
Cookville, TN 38502-2647

Knoxville Claims Association
P.O. Box 3609
Knoxville, TN 37922

Memphis Claims Association
5050 Poplar Ave., Suite 1130
Memphis, TN 38157

Nashville Claims Association
P.O. Box 17307
Nashville, TN 37217

Texas

Texas Independent Insurance
Adjusters Association
P.O. Box 709
Elgin, TX 78621

Texas Claims Association
P.O. Box 3928
Austin, TX 78764

Central Texas Claims
Association
P.O. Box 8908
Waco, TX 76714-8908

East Texas Area Claims
Association
P.O. Box 1328
Longview, TX 75606

North Texas Claims
Association
P.O. Box 2016
Wichita Falls, TX 76307

Pan Handle Claims
Association
P.O. Box 1750
Amarillo, TX 79105

Permian Basin Claims
Association
P.O. Box 9327
Odessa, TX 79105

Sabine-Nachez Claims
Association
87 Interstate 10 North
Suite 235
Beaumont, TX 77007

Valley Claims Association
P.O. Box 668
Harlingen, TX 78551

West Texas Claims Association
P.O. Box 2282
Abilene, TX 79604

Austin Claims Association
2550 S. Interstate Hwy 85
Suite 220
Austin, TX 78704

Corpus Christi Claims
Association
P.O. Box 8252
Corpus Christi, TX 78415

Dallas Claims Association
P.O. Box 660028
Dallas, TX 75266

El Paso Claims Association
P.O. Box 3157
El Paso, TX 79923

Ft. Worth Claims Association
1701 River Run, Suite 701
Ft. Worth, TX 76107

Houston Claims Association
600 Jefferson
Houston, TX 77002

San Angelo Claims
Association
P.O. Box 610
San Angelo, TX 76092

San Antonio Claims
Association
P.O. Box 290234
San Antonio, TX 78280-1634

Utah

Utah Claims Adjuster's
Association
5300 S. 360 West
Salt Lake City, UT 84102

Vermont

Vermont Claims Association
P.O. Box 456
Barre, VT 05641

Virginia

Virginia State Claims
Association
5148 Lakeshores Drive
Virginia Beach, VA 23455

Northern Virginia Claims
Association
P.O. Box 1324
Falls Church, VA 22041

Mid Valley Claims Association
P.O. Box 1261
Harrisonburg, VA 22801-1261

Penninsula Claims Association
P.O. Box 12129
Norfolk, VA 23502

Roanoke Claims Association
P.O. Box 7841
Roanoke, VA 24019

Washington

Washington State
Independent Adjusters
P.O. Box 87
Yakama, WA 98907

Seattle Claims Adjusters
Puget Power Building
OBC 6 South
Bellevue, WA 98009

Seattle Claims Manager's
Council
P.O. Box 88888
Seattle, WA 98188

Wenatchee Claims Adjuster's
Association
P.O. Box 1991
Wenatchee, WA 98801

West Virginia

Huntington Tri-State Claims
Association
P.O. Box 7804
Huntington, WV 25778-7804

Charleston Claims Association
P.O. Box 8551
So. Charleston, WV 25303

Wisconsin

Brown County Claims
Adjuster's Association
P.O. Box 13277
Green Bay, WI 54307-3277

Fox Valley Adjuster's
Association
711 N. Lynndale Dr.
Appleton, WI 54911

Wisconsin Claims Council
P.O. Box 539
Appleton, WI 54911

Milwaukee Claims Manager's
Council
P.O. Box 621
Milwaukee, WI 53201-53233

Milwaukee Insurance
Adjuster's Association
2201 Springdale Road
Waukesha, WI 53186

As previously mentioned, the state commissioner is a source of information for compiling a list of insurance agencies to contact. Address all correspondence to The Insurance Commissioner of (State).

Alabama
135 South Union St., Suite 160
Montgomery, AL 36130

Alaska
P.O. Box "D",
Juneau, AK 99811

Arizona
801 E. Jefferson, 2nd Floor,
Phoenix, AZ 85034

Arkansas
400 University Tower Bldg.
12th and University Streets
Little Rock, AR 72204

California
600 South Commonwealth
14th Floor
Los Angeles, CA 90005

Colorado
303 West Colfax Ave.
5th Floor
Denver, CO 80204

Connecticut
165 W. Capitol Ave.
State Office Bldg., Room 425
Hartford, CT 06106

Delaware
841 Silver Lake Blvd.
Dover, DE 19901

District of Columbia
614 "H" St., NW, Suite 512
Washington, DC 20001

Florida
State Capitol, Plaza Level 11
Tallahassee, FL 32399-0300

Georgia
2 Martin Luther King, Jr. Dr.
Floyd Memorial Bldg.
704 West Tower
Atlanta, GA 30334

Hawaii
P.O. Box 3614
Honolulu, HI 96811

Idaho
700 W. State Street
Boise, ID 83720

Illinois
320 W. Washington, 4th Floor
Springfield, IL 62767

Indiana
311 W. Washington St.
Suite 300
Indianapolis, IN 46204-2787

Iowa
Lucas State Office Bldg.
6th Floor
Des Moines, IA 50319

Kansas
420 SW 9th Street
Topeka, KS 66612

Kentucky
229 W. Main Street
Frankfort, KY 40602

Louisiana
P.O. Box 44214
Baton Rouge, LA 70804

Maine
State Office Bldg.
State House, Station 34
Augusta, ME 04333

Maryland
501 St. Paul Place, 7th Floor
Baltimore, MD 21202

Massachusetts
100 Cambridge St.
Boston, MA 02202

Michigan
P.O. Box 30220
Lansing, MI 48909

Minnesota
500 Metro Square Bldg.
5th Floor
St. Paul, MN 55101

Mississippi
1804 Walter Sillers Bldg.
Jackson, MS 39205

Missouri
301 W. High St., No. 6
Jefferson City, MO 65102-0690

Montana
126 North Sanders,
Mitchell Bldg., Room 270
Helena, MT 59601

Nebraska
301 Centennial Mall South
Lincoln, NE 68509

Nevada
Nye Bldg.
201 South Fall Street
Carson City, NV 89701

New Hampshire
169 Manchester
Concord, NH 03301

New Jersey
201 E. State St., CN325
Trenton, NJ 08625

New Mexico
PERA Bldg.
P.O. Drawer 1269
Santa Fe, NM 87504-1269

New York
160 West Broadway
New York, NY 10013

North Carolina
Dobbs Bldg., Box 26387
Raleigh, NC 27611

North Dakota
Capitol Bldg., 5th Floor
Bismarck, ND 58505

Ohio
2100 Stella Court
Columbus, OH 43266-0566

Oklahoma
P.O. Box 53408
Oklahoma City, OK
73152-3408

Oregon
158 12th Street, N.E.
Salem, OR 97310

Pennsylvania
Strawberry Square
Harrisburg, PA 17120

Rhode Island
100 North Main
Providence, RI 02903

South Carolina
1612 Marion Street
Columbia, SC 29202-3105

South Dakota
Insurance Bldg.
910 E. Sioux Ave.
Pierre, SD 57501

Tennessee
1808 West End Ave.
14th Floor
Nashville, TN 37219-5318

Texas
1110 San Jacinto Blvd.
Austin, TX 78701-1998

Utah
P.O. Box 45803
Salt Lake City, UT 84145

Vermont
State Office Bldg.
Montpelier, VT 05602

Virginia
700 Jefferson Bldg.
Richmond, VA 23209

Washington
Insurance Bldg. AQ21
Olympia, WA 98504

West Virginia
2100 Washington St. E.
Charleston, WV 25305

Wisconsin
123 W. Washington
Madison, WI 53707

Wyoming
Herschler Bldg.
122 West 25th Street
Cheyenne, WY 82002

A/C	Air Conditioner or part of air conditioning system (A/C duct, etc.)
Amp.	Ampere
Avg.	Average
Bdl.	Bundle (such as shingles)
B/F	Board Foot
BTU	British Thermal Unit
C/B	Concrete (cement) block
C/F	Cubic Foot
Dia.	Diameter
Ea.	Each
F.O.B.	Freight on Board (Free on Board)
Ga.	Garage
Hr.	Hour
Ht.	Height
Lb.	Pound
L/F	Linear Foot
M	Per one thousand
Max.	Maximum
Min.	Minimum
O/C	On Center
P.C.S.	Pieces
Pr.	Pair
S.F. or ⨍	Square Foot
Sq. Yd.	Square Yard
Std.	Standard
T & G	Tongue and Groove
W	Width

Symbols

%	Percent
"	Inches
'	Feet
x	By
#	Number
@	At
⨍ S.F.	Square Foot

The following pages contain worksheets for producing an individualized unit cost guide. The units listed are typical items repaired or replaced in insurance repair construction projects. There is a space for standard unit costs, which can be obtained from the current edition of published cost guides such as Means *Building Construction Cost Data*, or *Means Repair and Remodeling Cost Data*. There is also a space for local unit costs, which can be obtained from the contractor's historical cost data files or from local manufacturers. A contractor may not require this extensive listing and it is recommended that items be omitted which will not be typically used on local projects. Since unit cost guides do not reflect a contractor's overhead and profit, a percentage for overhead and profit should be added to the subtotal of the estimate.

Item	Unit	Standard Unit Cost	Local Unit Cost
Miscellaneous			
Volume Sand Blasting	S.F.		
Steam Pressure Cleaning	S.F.		
Roof Costing	S.F.		
Exterior Cleaning – window washing	S.F.		
Area deodorizing	S.F.		
Galvanized Termite Shield	S.F.		
Interior Shutters (rule of thumb)	S.F.		
Air Conditioning			
Window Units			
6,000 B.T.U. – 1/2 ton	Ea.		
9,000 B.T.U. – 3/4 ton	Ea.		
12,000 B.T.U. – 1 ton	Ea.		
18,000 B.T.U. – 1-1/2 ton	Ea.		
24,000 B.T.U. – 2 ton	Ea.		
Central Units (complete with ducts)			
To 500 S.F. of floor area – 1 ton	Ea.		
To 501–1,000 S.F. of floor area – 2 ton	Ea.		
To 1,001–1,600 S.F. of floor area – 3 ton	Ea.		
Awnings (window/door, not patio/porch covers)			
Aluminum	S.F.		
Fiberglass (synthetic materials)	Sq. Yd.		
Cloth (recover frames)	Sq. Yd.		
Ventilated (lath type) multiple slat lath	S.F.		
add for fold down (storm type)	S.F.		

Item	Unit	Standard Unit Cost	Local Unit Cost
Bathroom Fixtures & Accessories			
Miscellaneous			
Cup & Brush Holder (chrome)	Ea.	_____	_____
" " " (ceramic)	Ea.	_____	_____
" " " (plastic)	Ea.	_____	_____
Grab Bars – 24" (towel holder)	Ea.	_____	_____
" " " – "L" shaped – 16" x 32"	Ea.	_____	_____
Bath Exhaust Fan (Complete, including wiring)	Ea.	_____	_____
Medicine Cabinets			
Above surface unit – swing door – 36" x 28"	Ea.	_____	_____
(adjust price by size accordingly)		_____	_____
Recessed unit – swing door			
16" x 26"	Ea.	_____	_____
16" x 26" with light	Ea.	_____	_____
16" x 28" with side light	Ea.	_____	_____
Recessed with bypass (sliding doors)			
2 doors – 14" x 18" with top light	Ea.	_____	_____
" " – 14" x 18" with side lights	Ea.	_____	_____
Bath Vanity Cabinets (Formica tops, simple or modern design. Includes plumbing connection, not fixtures.)			
24" width	Ea.	_____	_____
30" width	Ea.	_____	_____
36" width	Ea.	_____	_____
add for fancy raised design			
24"	Ea.	_____	_____
30"	Ea.	_____	_____
36"	Ea.	_____	_____

Item	Unit	Standard Unit Cost	Local Unit Cost
Bathroom Fixtures & Accessories (continued)			
Shower and Tub Doors			
(double door sliding)			
Hinged 60″ x 28″	Ea.		
Sliding 70″ x 60″	Ea.		
Sliding 60″ x 60″	Ea.		
Cabinets (see page 2 for bathroom cabinets)			
Kitchen Cabinets:			
Lower B Cabinets $_____L.F.	L.F.		
Upper Cabinets $_____L.F.	L.F.		
(plus hourly labor estimated for removal and installation)			
Economy Units (no tops, plywood or composition drawers, exposed hinges)			
Base Unit – $_____ per unit	L.F.		
Wall Unit – $_____ per unit	L.F.		
Better Quality (avg. middle class home; hidden hinges)			
Base Unit – $_____ per unit	L.F.		
Wall Unit – $_____ per unit	L.F.		
Quality (upper class homes; hidden hinges, magnetic catches, solid core doors)			
Base Unit – $_____ per unit	L.F.		
Wall Unit – $_____ per unit	L.F.		
Clean and wax exterior			
Clean interior			
Refinish exterior surface	S.F.		
Broom closet complete			
Corner cabinets; plain 24″ x 24″ x 30″			
″　　″ (revolving shelves) 24″ x 24″ x30″			

Item	Unit	Standard Unit Cost	Local Unit Cost
Cabinets (continued)			
Oven cabinet – 29″ x 29″ x 84″			
Window valance	L.F.		
Refinish inside and outside complete, including R & R and spray doors: (inside 4 times surface area of outside)	S.F.		
Canopies/Carports (patio, porch and carport covers)			
Aluminum (painted) – cover only	S.F.		
″ ″ ″ – complete			
″ ″ ″ – embossed cover only	S.F.		
″ ″ ″ – embossed complete			
″ ″ ″ – supports	L.F.		
Top and bottom plates	L.F.		
Corner posts			
Fiberglass	S.F.		
Screening	S.F.		
″ ″ – overhead	S.F.		
Aluminum			
Screening			
″ ″ – overhead	S.F.		
Ceilings – Ceiling Tile			
Acoustical ceiling tile nailed, stapled or clipped in place			
Cushion Tone 12″ x 12″ T & G	S.F.		
″ ″ 12″ x 24″	S.F.		
Celotex 12″ x 12″	S.F.		
Non-acoustical (plain)			
12″ x 12″ T & G	S.F.		
12″ x 24″ T & G	S.F.		
12″ x 12″ painted	S.F.		
Removal to save furring strips			

Item	Unit	Standard Unit Cost	Local Unit Cost
Ceilings – Ceiling Tile (continued)			
Furring strips			
1″ x 2″	S.F.	_____	_____
1″ x 3″	S.F.	_____	_____
1″ x 4″	S.F.	_____	_____
Suspended Ceilings (completely installed)	S.F.	_____	_____
Complete suspended ceiling system, grid work and tile (everything)			
Remove to save grid system		_____	_____
Cushion Tone 2′ x 4′		_____	_____
Fashion Tone 2′ x 4′		_____	_____
Temlock 2′ x 4′		_____	_____
Add for 2′ x 2′ panels		_____	_____
Add for scaffolding, if needed		_____	_____
Drywall Ceiling			
Minimum charge, incl. taping & sanding	2 Trips	_____	_____
Replace 4′ x 8′ x 1/2″ sheet rock only	S.F.	_____	_____
Replace and finish complete 3/8″	S.F.	_____	_____
1/2″ complete	S.F.	_____	_____
Finish only	S.F.	_____	_____
Plastered Ceilings			
Minimum charge	1 Trip	_____	_____
	2 Trips	_____	_____
Under 50 sq. yd.	S.Y.	_____	_____
Removal	S.Y.	_____	_____
Over 50 sq. yd.	S.Y.	_____	_____
Over 1,000 sq. yd.	S.Y.	_____	_____
Add for swirl finish	S.Y.	_____	_____
Add decorative – cove	S.Y.	_____	_____

Item	Unit	Standard Unit Cost	Local Unit Cost
Plastered Ceilings (continued)			
Add simulated blown on, stippled or sparkle	S.F.		
Concrete Products			
Concrete blocks (all sizes)	S.F.		
4" x 8" x 16"			
8" x 8" x 16" (most common)			
12" x 8" x 16"			
8" x 4" x 16"			
Minimum charge			
Brick			
Common	S.F.		
Face brick (depending on kind, style, type, size, etc.)	S.F.		
Slump brick	S.F.		
Lintel			
Poured reinforced	L.F.		
8" x 8" precast	L.F.		
minimum charge (for poured)			
Removal of block or brick (complete demolition)	S.F.		
Tear out for piecework	S.F.		
Stone			
Coral rock rubble material	S.F.		
Keystone, random or ash	S.F.		
Concrete Work			
Sidewalk – walkway 4"	S.F.		
Driveway – 6"	S.F.		
Curb and gutter	L.F.		
Cap Block 4" x 16"	L.F.		
6" x 16"	L.F.		
8" x 16"	L.F.		

Item	Unit	Standard Unit Cost	Local Unit Cost
Concrete Products (continued)			
Ceramic Tile			
Floor or wall 4¼″ x 4¼″	S.F.	_____	_____
glazed 4″ x 4″ (rule of thumb) including walls and/or floor)	S.F.	_____	_____
Cove base or corners	L.F.	_____	_____
Floor 4¼″ x 4¼″	S.F.	_____	_____
Counter Tops			
Formica			
Counter top 4″ backsplash	L.F.	_____	_____
Plain – no cutouts	L.F.	_____	_____
With 18″ backsplash	L.F.	_____	_____
Factory pre-formed 4″–6″ backsplash	L.F.	_____	_____
Furnish & install stainless steel pan		_____	_____
Cutting board or ceramic top filler		_____	_____
Plus add R & R sink	Ea.	_____	_____
R & R sink with garbage disposal	Ea.	_____	_____
R & R cook top surface unit	Ea.	_____	_____
Add for corner angles	Ea.	_____	_____
Add for miter corner	Ea.	_____	_____
Consider appearance allowances, cutting boards, ceramic filler, etc.			
Debris Removal – Tear Out			
Tear Out			
Foreman (usually not needed)	Per Hr.	_____	_____
Skilled laborer	Per Hr.	_____	_____
Unskilled laborer	Per Hr.	_____	_____
Debris Removal			
Small jobs (1/2- to 3/4-ton pick-up truck, driver and helper, including dump fee)	Ea. Load	_____	_____

Item	Unit	Standard Unit Cost	Local Unit Cost
Debris Removal – Tear Out (continued)			
Large jobs (1- to 1½-ton dump truck, driver, 1 or 2 helpers, including dump fee)	Ea. Load	_____	_____
Deodorizing	S.F.	_____	_____
Doors			
Exterior Doors			
Jalousie – complete with jamb, hardware, trim and glass			
3' 0" x 6' 8" x 1¾"	Ea.	_____	_____
2' 8" x 6' 8" x 1¾"	Ea.	_____	_____
2' 6" x 6' 8" x 1¾"	Ea.	_____	_____
Jalousie – door only, use existing hardware and trim			
3' 0" x 6' 8" x 1¾"	Ea.	_____	_____
2' 8" x 6' 8" x 1¾"	Ea.	_____	_____
2' 6" x 6' 8" x 1¾"	Ea.	_____	_____
Solid Core – complete with jamb, brickmold and trim			
3' 0" x 6' 8" x 1¾"	Ea.	_____	_____
2' 8" x 6' 8" x 1¾"	Ea.	_____	_____
2' 6" x 6' 8" x 1¾"	Ea.	_____	_____
Solid Core – door only, use existing hardware and trim			
3' 0" x 6' 8" x 1¾"	Ea.	_____	_____
2' 8" x 6' 8" x 1¾"	Ea.	_____	_____
2' 6" x 6' 8" x 1¾"	Ea.	_____	_____
Hollow Core – complete with jamb, brickmold and trim			
3' 0" x 6' 8" x 1¾"	Ea.	_____	_____
2' 8" x 6' 8" x 1¾"	Ea.	_____	_____
2' 6" x 6' 8" x 1¾"	Ea.	_____	_____

Item	Unit	Standard Unit Cost	Local Unit Cost
Doors (continued)			
Hollow Core – door only, use existing hardware and trim			
3′ 0″ x 6′ 8″ x 1¾″	Ea.		
2′ 8″ x 6′ 8″ x 1¾″	Ea.		
2′ 6″ x 6′ 8″ x 1¾″	Ea.		
Add for hardware (per door)	Ea.		
Add for closers (per door)	Ea.		
Special mahogany or teak doors, single or double, carved or moulded, on quotation only			
Pre-package Units – pre-hung, including door, jamb, stops, casing & hardware, assembled			
2′ 0″ x 6′ 8″ x 1⅜″	Ea.		
2′ 6″ x 6′ 8″ x 1⅜″	Ea.		
2′ 8″ x 6′ 8″ x 1⅜″	Ea.		
3′ 0″ x 6′ 8″ x 1⅜″	Ea.		
Add for solid core interior doors			
Folding interior doors – nylon, vinyl or other plastic for 6′ 8″ high openings			
3′ wide	Ea.		
5′ wide	Ea.		
6′ wide	Ea.		
8′ wide	Ea.		
10′ wide	Ea.		
12′ wide	Ea.		
Sliding glass doors (interior–exterior); tempered glass doors with aluminum frames and tracks			
6′ x 6′ 8″ – 2 panel″	Ea.		
8′ x 6′ 8″ – 2 panel″	Ea.		

Item	Unit	Standard Unit Cost	Local Unit Cost
Doors (continued)			
9' x 6' 8" – 3 panel	Ea.		
10' x 6' 8" – 2 panel	Ea.		
Screens for sliding doors – complete			
3' 0" x 6' 8"	Ea.		
4' 0" x 6' 8"	Ea.		
5' 0" x 6' 8"	Ea.		
screening only	S.F.		
Interior doors, flush hollow-core, complete with hardware			
3' 0" x 6' 8" x 1⅜"	Ea.		
2' 8" x 6' 8" x 1⅜	Ea.		
2' 6" x 6' 8" x 1⅜	Ea.		
Interior doors, flush hollow-core, door only, installed			
3' 0" x 6' 8" x 1⅜"	Ea.		
2' 8" x 6' 8" x 1⅜	Ea.		
2' 6" x 6' 8" x 1⅜	Ea.		
Add for interior lockset and hinges, if required	Set		
Cafe Type Doors (usually found in kitchens, swing type)			
3' 0" x 3' 8"	Ea.		
2' 8" x 3' 8"	Ea.		
2' 6" x 3' 6"	Ea.		
Add for gravity hinges (2 pr/set)	Set		
Screen Doors			
3' 0" x 6' 8" x 1⅛" – 2 panel wood, complete with hardware	Ea.		
3' 0" x 6' 8" x 1⅛" – 2 panel wood, door only	Ea.		

Item	Unit	Standard Unit Cost	Local Unit Cost
Doors (continued)			
Add for closer, installed (Smaller size doors same price. 7′ screen doors available by special order only.)	Ea.	_____	_____
Pocket Doors – complete with jamb, pocket, hardware & trim			
3′ 0″ x 6′ 8″	Ea.	_____	_____
2′ 8″ x 6′ 8″	Ea.	_____	_____
2′ 6″ x 6′ 8″	Ea.	_____	_____
By-Pass Closet Doors – 2 doors, complete with jamb, casing, hardware and tracks			
36″ opening	Ea.	_____	_____
48″ opening	Ea.	_____	_____
60″ opening	Ea.	_____	_____
72″ opening	Ea.	_____	_____
Garage Doors (All garage doors are listed as installed complete on existing jambs.) Wood Doors – no windows, installed			
8′ x 7′ (2 panels, 4 sections)	Ea.	_____	_____
9′ x 7′ (3 panels, 4 sections)	Ea.	_____	_____
10′ x 7′ (4 panels, 3 sections) (heavy)	Ea.	_____	_____
16′ x 7′ (4 panels, 4 sections)	Ea.	_____	_____
18′ x 7′ (5 panels, 4 sections)	Ea.	_____	_____
Add for windows (per door)		_____	_____
Aluminum Doors – no windows, installed "sectional"			
8′ x 7′	Ea.	_____	_____
9′ x 7′	Ea.	_____	_____
10′ x 7′	Ea.	_____	_____
16′ x 7′	Ea.	_____	_____
18′ x 7′	Ea.	_____	_____

Item		Unit	Standard Unit Cost	Local Unit Cost
Doors (continued)				
Steel Doors – residential				
8′ x 7′	24 GA. Flush	Ea.	_____	_____
9′ x 9′	″ ″ ″	Ea.	_____	_____
16′ x 7′	″ ″ ″	Ea.	_____	_____
16′ x 8′	Special	Ea.	_____	_____
18′ x 7′	″	Ea.	_____	_____
Steel Doors – commercial				
10′ x 10′	Standard Steel	Ea.	_____	_____
10′ x 12′	″ ″ ″	Ea.	_____	_____
12′ x 14′	″ ″ ″	Ea.	_____	_____
Add for new jambs, if required (per door)				
Miscellaneous Door Parts				
Add for door jamb (frame) and door trim		Ea.	_____	_____
Add for door trim		Per Side	_____	_____
Door lock and handle installed (lockset)		Ea.	_____	_____
Hinges (installed)		Pr.	_____	_____
Screen door lock		Ea.	_____	_____
Screen door hinges		Pr.	_____	_____
Average door knob		Ea.	_____	_____
Door chimes, electric, 2-tone		Ea.	_____	_____
″ ″ ″ non-electric		Ea.	_____	_____
Downspouts				
Galvanized 3″		L.F.	_____	_____
Galvanized 4″		L.F.	_____	_____
T and L joint fasteners		Ea.	_____	_____

Item	Unit	Standard Unit Cost	Local Unit Cost
Downspouts (continued)			
Copper 3"	L.F.		
Copper 4"	L.F.		
Dry Wall (Sheetrock)			
Sheet rock 3/8" x 4' x 8'	S.F.		
Add for 1/2"	S.F.		
Minimum charge, incl. topping, sanding	2 Trips		
Replace 4' x 8' x 3/8" sheetrock only	S.F.		
" " " 1/2" " " "	S.F.		
Replace and finish complete – 3/8"	S.F.		
" " " " " " – 1/2"	S.F.		
Finish only over float & tape existing work	S.F.		
Electrical (Guidelines only)			
Circuit breaker installation	Ea.		
100 amp. 8 spaces	Ea.		
100 amp. 10 spaces	Ea.		
Dishwasher – new installation	Ea.		
Labor and material	Ea.		
Dryer receptacle, incl. connect in box	Ea.		
Wiring for door bells or chimes	Ea.		
Garbage disposal	Ea.		
Connect only	Ea.		
Heater	Ea.		
Range	Ea.		
Thermostat	Ea.		
Water Heater	Ea.		
Connect only	Ea.		

Item	Unit	Standard Unit Cost	Local Unit Cost
Electrical (continued)			
Washer and dryer together	Set		
Wiring per outlet	Ea.		
Wiring per wall switch – single	Ea.		
″ ″ ″ ″ ″ – double	Ea.		
Stove hood with fan	Ea.		
Stove hood fan (new work from box out)	Ea.		
Outside post light	Ea.		
Pool pump & light (new work from box out)	Set		
Central A/C system	Ea.		
100 amp. service panel	Ea.		
150 amp. service panel	Ea.		
200 amp. service panel	Ea.		
Rewire old R & R – per outlet	Ea.		
Temporary power (pole, meter, outlet)	Set		
Service weatherhead with pole			
100 amp.	Ea.		
150 amp.	Ea.		
200 amp.	Ea.		
Add for each L.F. over 6′ (pole ht.)	L.F.		
Floor Coverings			
Carpet			
Remove carpets and replace	S.F.		
Installation – new carpet	S.Y.		
Shampoo carpet (clean)	S.F.		
Beat carpet to remove glass	S.F.		
Add for shag	S.F.		

Item	Unit	Standard Unit Cost	Local Unit Cost
Floor Coverings (continued)			
Terrazzo (new)			
New installation	S.F.		
Grind, seal, polish (stained, scorched)	S.F.		
Strip, seal, polish	S.F.		
Clean, polish	S.F.		
Minimum charge for repairs (patching, fill cracks, etc.)	Job		
Parquet			
Parquet blocks to 150 S.F.	S.F.		
" " " 150 to 300 S.F.	S.F.		
" " " over 300 S.F.	S.F.		
Remove old to prepare floor surface for new	S.F.		
Add 10–15% for waste and matching to actual covered area (minimum charge for any repairs)	Job		
If required, add for R & R shoe moulding	L.F.		
Hardwood Floors			
T & G Pine 1" x 4"	S.F.		
T & G Pine 1" x 3"	S.F.		
Ash or Oak 1½" x 2½"	S.F.		
Minimum repair charge	Job		
Clean and wax	S.F.		
Sand and refinish	S.F.		
Remove & prepare surface for refinishing	S.F.		
Miscellaneous Floor Covering (Depending on type, design, thickness, style, etc.)			
Asphalt tile	S.F.		
Vinyl asbestos	S.F.		
Rubber	S.F.		

Item	Unit	Standard Unit Cost	Local Unit Cost
Miscellaneous Floor Coverings			
Pure vinyl	S.F.		
Sheet vinyl – 6′ (secure quote)	S.F.		
Linoleum	S.F.		
Masonite underlayment	S.F.		
Plywood underlayment	S.F.		
Sand, refinish and seal floor	S.F.		
Floor Mouldings			
Base moulding	L.F.		
Shoe ″	L.F.		
Crown ″	L.F.		
Add per room, incl. labor & material	Room		
1/4″ round moulding	L.F.		
Cove moulding	L.F.		
Paint floor moulding (if enamel) coat concrete	L.F.		
Preparation additional	S.F.		
Glass and Glazing			
Rule of thumb	S.F.		
Single strength door & window panes	S.F.		
Double ″ ″ ″ ″	S.F.		
Crystal single strength	S.F.		
″ double ″	S.F.		
Plate glass			
Allow for solar and colored glass	S.F.		
Mirrors (rule of thumb)	S.F.		

Item	Unit	Standard Unit Cost	Local Unit Cost
Gutters and Downspouts			
Galvanized 3″ diameter	L.F.		
″ ″ 4″ ″	L.F.		
Copper 3″ ″	L.F.		
″ ″ 4″ ″	L.F.		
T's, L's and fasteners	Ea.		
Galvanized gutter 2″ x 4″	L.F.		
Insulation			
Rock wool batts – 6″	S.F.		
Fiberglass batts – 4″ or equivalent	S.F.		
Blown insulation – 4″ or equivalent	S.F.		
Fiberglass batts – 2″ or equivalent	S.F.		
Poured fiberglass – 3″ or equivalent	S.F.		
Aluminum foil – 2″ backed	S.F.		
Aluminum foil – 3″ backed	S.F.		
Asphalt sheathing, temlock black fiberboard	S.F.		
Lumber			
(Framing lumber: 2″ x 4″, 2″ x 6″, 2″ x 8″, 4″ x 4″, 1″ x 12″, 1″ x 4″, etc.)			
Normal size jobs	B.F.		
Under 1,000	B.F.		
Decking & sheathing subfloor, incl. waste	S.F.		
Removal decking & sheathing	S.F.		
1″ x 3″ cross bridging or 2″ solid	L.F.		
1″ x 3″ nailing strips	L.F.		
Avg. labor for woodwork	B.F.		
Avg. materials	B.F.		

Item	Unit	Standard Unit Cost	Local Unit Cost
Painting			
Exterior Painting	S.F.	_____	_____
On stucco, cement block, brick, or concrete – 1 coat	S.F.	_____	_____
All vinyl, water, cement base – 2 coats	S.F.	_____	_____
Exterior on wood			
Enamel – 1 coat, incl. normal sanding	S.F.	_____	_____
" " – 2 coats	S.F.	_____	_____
Exterior openings (doors and windows)			
Windows – enamel, 1 coat	Per Side	_____	_____
" " " " 2 coats	Per Side	_____	_____
Doors – enamel, 1 coat	Per Side	_____	_____
" " " " 2 coats	Per Side	_____	_____
Screen Door – enamel, 1 coat	Per Side	_____	_____
" " " " 2 coats	Per Side	_____	_____
Shutters – enamel, 1 coat	Per Side	_____	_____
" " " " 2 coats	Per Side	_____	_____
Window screens – enamel, 1 coat	Per Side	_____	_____
Nailing strips on screening	L.F.	_____	_____
Paint complete cornice (soffit & fascia)	S.F.	_____	_____
Interior Painting			
Clean and paint, 1 coat (latex, Kemtone, etc.)	S.F.	_____	_____
Paint only, no cleaning at all	S.F.	_____	_____
Cleaning & painting, incl. cleaning of windows and screens	S.F.	_____	_____
Clean and paint, 1 coat – enamel	S.F.	_____	_____
" " " 2 coats "	S.F.	_____	_____
Clean and paint, 1 coat, incl. cleaning of windows and screens	S.F.	_____	_____

Item	Unit	Standard Unit Cost	Local Unit Cost
Painting (continued)			
Painting only, enamel or latex – no window cleaning or screen cleaning	S.F.		
Kitchen Cabinets & Miscellaneous			
Minimum charge for painting	Job		
Spray paint cabinets, including sanding, puttying, priming, sealing and refinishing complete	S.F.		
Interior S.F. = 4 times exterior measure Brush paint same as above			
Interior and exterior complete	S.F.		
Exterior only, 1 coat	S.F.		
" " " 2 coats	S.F.		
Note: Unit cost for kitchen cabinets includes R & R doors, hinges, hardware			
Spray paint refrigerator	Ea.		
Paint closet trim complete, 1 shelf, 2 sides, hang rail and bar			
Note: All interior room painting, add per room for trim painting			
Brush paint sparkle over acoustical blown ceiling	S.F.		
Plumbing			
Water Heaters – gas or electric 30 gallon capacity	Ea.		
40 " " "	Ea.		
50 " " "	Ea.		
Labor Rate, truck and 1 plumber	Per Hr.		
" " " " " 2 plumbers	Per Hr.		
Minimum charge, truck and 1 plumber	1 Call		
" " " " "	2 Calls		

Item	Unit	Standard Unit Cost	Local Unit Cost
Plastering			
Minimum charge	1 Trip	_____	_____
	2 Trips	_____	_____
Under 50 S.Y.	S.Y.	_____	_____
Removal	S.Y.	_____	_____
Over 50 S.Y.	S.Y.	_____	_____
Over 1,000 S.Y.	S.Y.	_____	_____
Add for swirl finish	S.Y.	_____	_____
Add decorative – cove	S.Y.	_____	_____
Add simulated blown on, stippled or sparkle	S.Y.	_____	_____
Ranges, Hoods, and Power (Fan) Units			
Built-in Ovens – Standard	Ea.	_____	_____
Better Quality with window		_____	_____
Add for rotisserie		_____	_____
Ductless Range Hoods			
Colored 30″	Ea.	_____	_____
36″	Ea.	_____	_____
42″	Ea.	_____	_____
Anodized Aluminum 30″	Ea.	_____	_____
36″	Ea.	_____	_____
42″	Ea.	_____	_____
Stainless Steel 30″	Ea.	_____	_____
36″	Ea.	_____	_____
42″	Ea.	_____	_____
Installation labor	Ea.	_____	_____

Item	Unit	Standard Unit Cost	Local Unit Cost
Ranges, Hoods, & Power (Fan) Units (continued)			
Vented Range Hoods (with duct)			
Colored 30″	Ea.		
36″	Ea.		
42″	Ea.		
Anodized Aluminum 30″	Ea.		
36″	Ea.		
42″	Ea.		
Stainless Steel 30″	Ea.		
36″	Ea.		
42″	Ea.		
Installation labor (all above using Nutone or equivalent)			
Standard installation	Ea.		
Install deluxe or 3-speed blowers & fan	Ea.		
Installation with vented roof	Ea.		
Backsplash Exhaust Units			
Colored 30″	Ea.		
36″	Ea.		
42″	Ea.		
Anodized Aluminum 30″	Ea.		
36″	Ea.		
42″	Ea.		
Stainless Steel 30″	Ea.		
36″	Ea.		
42″	Ea.		
Electrical connection	Ea.		

Item	Unit	Standard Unit Cost	Local Unit Cost
Ranges, Hoods, & Power (Fan) Units (continued)			
Range Top – Standard	Ea.	_____	_____
Better Quality	Ea.	_____	_____
Roofing			
Note: 1 square = 100 S.F.			
Note: Removal and hauling are included in all of the following prices. If not required, then deduct:	S.F.	_____	_____
Note: For buildings over 1 story, add to all following prices:	Square	_____	_____
Built-up 3-ply			
Minimum charge to patch	Job	_____	_____
" " " for complete replacement	Job	_____	_____
2–5 squares (3 ply gravel)	Square	_____	_____
6–10 " " "	Square	_____	_____
11–20 " " "	Square	_____	_____
Over 20 " " "	Square	_____	_____
Sweep, tar and regravel	Square	_____	_____
" " " " "	Min. Chg.	_____	_____
Add for red river rock	Square	_____	_____
Add for white marble chip	Square	_____	_____
Flat White Cement Tile			
Minimum charge to patch (open)	Job	_____	_____
" " " " for complete replacement	Job	_____	_____
2–5 squares	Square	_____	_____
over 5 squares	Square	_____	_____
Barrel Tile, White Cement			
Minimum charge to patch (open)	Job	_____	_____
" " " for complete replacement	Job	_____	_____

Item	Unit	Standard Unit Cost	Local Unit Cost
Roofing (continued)			
2–5 squares	S.F.		
over 5 squares	S.F.		
Other Types of Roofing			
Spanish Tile – special quote – not stocked			
Asphalt shingles (235#)	Square		
90# slate	Square		
Weather Vanes			
26″	Ea.		
30″	Ea.		
Cupolas			
(Natural finish aluminum roof)			
24″ x 24″ x 25″ high	Ea.		
30″ x 30″ x 28″ high	Ea.		
36″ x 36″ x 32″ high	Ea.		
add for weather vane, if applicable	Ea.		
Screening			
Fiberglass	S.F.		
Aluminum	S.F.		
Overhead	S.F.		
Paint wood fastening strip	L.F.		
Siding			
Asbestos	Square		
Shingles	Square		
Building paper	S.F.		
Minimum charge, incl. waste & matching	Job		
No. 1 wood	S.F.		

Item	Unit	Standard Unit Cost	Local Unit Cost
Siding (continued)			
C & B (CD) or equivalent	S.F.		
Removal	S.F.		
1/2″ gypsum	S.F.		
3/4″ waterproof celotex	S.F.		
1″ x 4″ batten strips	L.F.		
1″ x 3″ nailing strips	L.F.		
Aluminum			
8″ exposure	Square		
4″ exposure double lap	Square		
Moulding – per 12′ section			
Window & door trim – per 10′			
Outside corners – per 9′			
Inside corners – per 10′			
Starter strips – per 10′			
Soffit & fascia trim – per L.F.	L.F.		
Sills			
Marble	L.F.		
Ceramic	L.F.		
Stairs – Folding – Pulldown			
84″ x 105″			
105″ x 120″			
Stucco			
Tool out and patch – minimum charge	1 Trip		
″ ″ ″ ″ ″ ″ ″	2 Trips		
Removal	S.Y.		

Item	Unit	Standard Unit Cost	Local Unit Cost
Stucco (continued)			
Replace under 50 yds.	S.Y.	_____	_____
" over 50–100 yds.	S.Y.	_____	_____
" over 100 yds.	S.Y.	_____	_____
Blown on white coat stucco with marble chips	S.F.	_____	_____
Minimum charge for Perma Freeze (depending on quantity)	Job	_____	_____
Note: deduct for openings, add for masking off.			
Tile			
Note: Use exact room measurements. Prices below already include waste.			
Ceramic Tile			
Floor and wall 4¼″ x 4¼″	S.F.	_____	_____
Glazed 4″ x 4″	S.F.	_____	_____
Cove, base, corners	S.F.	_____	_____
1″ x 1″ floor tile	S.F.	_____	_____
Tub and shower enclosure complete (3 walls, soap dish, grab bar, towel holder)			
Aluminum tile	S.F.	_____	_____
Plastic tile	S.F.	_____	_____
Stainless steel	S.F.	_____	_____
Removal (all types)	S.F.	_____	_____
Underlayments (Roof sheathing, sub flooring, and wall sheathing)			
Asphalt 3/8″ wall sheathing	S.F.	_____	_____
" " 1/2″ " "	S.F.	_____	_____

Plywood sheathing

Size	Type	Grade			
3/8″ – 5/6″	Int.	CD	S.F.	_____	_____
1/2″ – 5/8″	Int.	CD	S.F.	_____	_____

Item			Unit	Standard Unit Cost	Local Unit Cost
Plywood Sheathing (continued)					
Size	*Type*	*Grade*			
3/8" – 5/16"	Int.	CD	S.F.		
1/2" – 3/4"	Int.	A/C	S.F.		
1/4" – 3/8"	Ext.	A/C	S.F.		
1/2" – 3/4"	Ext.	A/C	S.F.		
Masonite 1/8"			S.F.		
1/4"			S.F.		
1/4" interior plywood (all with 1 good side)			S.F.		
3/8" ″ ″ ″ ″ ″ ″ ″			S.F.		
1/2" ″ ″ ″ ″ ″ ″ ″			S.F.		
5/8" ″ ″ ″ ″ ″ ″ ″			S.F.		
3/4" ″ ″ ″ ″ ″ ″ ″			S.F.		
Roofing felt – 30 lb. roll, in place			Roll		
Wallpaper (Replacement prices include lap joints, butt joints, waste and matching)					
Clean Wallpaper			S.F.		
Removal Dry			S.F.		
Steam			S.F.		
Installation (labor and materials) Standard solid colors (no designs) Lap work			S.F.		
Butt work			S.F.		
Matching designs and patterns Lap work			S.F.		
Butt work			S.F.		

Item	Unit	Standard Unit Cost	Local Unit Cost
Wallpaper (continued)			
Vinyl coated colors and patterns		_____	_____
Lap work	S.F.	_____	_____
Butt work	S.F.	_____	_____
Vinyl flocked raised patterns			
Butt work	S.F.	_____	_____
Bordering			
Lap work	L.F.	_____	_____
Butt work	L.F.	_____	_____
Add for foils, grass cloth or silk (per roll)	Roll	_____	_____

Windows & Window Treatments
(Window prices do not include tear-out, preparation and repair of jambs or replacement of broken sills or damaged plaster and stucco.)

Item	Unit	Standard Unit Cost	Local Unit Cost
Vinyl Window Shades			
4'	Ea.	_____	_____
5'	Ea.	_____	_____
6'	Ea.	_____	_____
7'	Ea.	_____	_____
Wood Single/Double Hung Windows			
Average width 24"	Ea.	_____	_____
" " " 36"	Ea.	_____	_____
Single hung – average	Ea.	_____	_____
Sliding hung – 2 & 3 lite	Ea.	_____	_____

Awning Type Windows

Trade #	Width	Height	Unit	Standard Unit Cost	Local Unit Cost
1/2 3–2	26"	26"	Ea.	_____	_____
1/2 3–3	26"	38"	Ea.	_____	_____
1/2 3–4	26"	50"	Ea.	_____	_____
2–2	37"	26"	Ea.	_____	_____

Item			Unit	Standard Unit Cost	Local Unit Cost
Awning Type Windows (continued)					
Trade #	*Width*	*Height*			
2–3	37"	38"	Ea.	_____	_____
2–4	38"	50"	Ea.	_____	_____
2–5	37"	62"	Ea.	_____	_____
3–2	54"	26"	Ea.	_____	_____
3–3	54"	38"	Ea.	_____	_____
3–4	54"	50"	Ea.	_____	_____
3–5	54"	62"	Ea.	_____	_____
Jalousie Type Windows					
Trade #	*Width*	*Height*			
1/2 2–2	19"	26"	Ea.	_____	_____
1/2 2–3	19"	38"	Ea.	_____	_____
1/2 2–4	19"	50"	Ea.	_____	_____
1/2 2–5	19"	62"	Ea.	_____	_____
1/2 3–2	26"	26"	Ea.	_____	_____
1/2 3–3	26"	38"	Ea.	_____	_____
1/2 3–4	26"	50"	Ea.	_____	_____
2–2	37"	26"	Ea.	_____	_____
2–3	37"	38"	Ea.	_____	_____
2–4	37"	50"	Ea.	_____	_____
2–5	37"	62"	Ea.	_____	_____
3–2	54"	26"	Ea.	_____	_____
3–3	54"	38"	Ea.	_____	_____
3–4	54"	50"	Ea.	_____	_____
3–5	54"	62"	Ea.	_____	_____

Skill	Standard Hourly Wage	Local Hourly Wage
Foreman	_____	_____
Skilled laborer (craftsman)	_____	_____
Semi-skilled (journeyman)	_____	_____
Unskilled (laborer)	_____	_____
Mason (bricklayer – stone)	_____	_____
Minimum Charge	_____	_____
Plasterer/Drywall, min. charge 1 trip	_____	_____
" " " " " " 2 trips	_____	_____
Plumber One truck and man	_____	_____
Journeyman & trainee (2 men)	_____	_____
Minimum charge (1 man/truck) 1 call	_____	_____
" " " " " " 2 calls	_____	_____
Electrician One truck and man	_____	_____
Minimum charge, 1 call	_____	_____
" " " " 2 calls	_____	_____
Electrician & helper	_____	_____

Item	Unit	Standard Unit Cost	Local Unit Cost
Carpet			
Wall-to-Wall Carpet Cleaning			
Basic charge for normal cleaning	S.F.		
Add for heavy soil, shampoo residue or unusual circumstances	S.F.		
Minimum charge	Job		
Area Rugs			
Basic charge for normal cleaning	S.F.		
Add for excess soil, shampoo, etc.	S.F.		
Reversibles	S.F.		
Hook rugs	S.F.		
Oriental or exotic rugs	S.F.		
Pick up and delivery ($15.00 min. each way)	S.F.		
Take up and reinstallation	S.F.		
Storage (Cost per week per item)			
Repairs		Quote	
Emergency Spotting Service			
First 1/2 hour	S.F.		
Each additional half hour	S.F.		
Special Treatments			
Deodorization (chemical)	S.F.		
Soil and stain repellent	S.F.		
Anti-static	S.F.		
Sanitizing (after septic contamination)	S.F.		
Mildewcide	S.F.		
Water Damage Restoration			
Emergency water extraction	S.F.		
Disengage carpet & remove pad	S.F.		

Item	Unit	Standard Unit Cost	Local Unit Cost
Carpet (continued)			
Mildew treatment	S.F.		
Sanitization (after septic contamination)	S.F.		
Cartage and disposal of pad and/or carpet as required			
Supervisor:	Hr.		
Assistant:	Hr.		
Drying equipment (per unit)	Day		
Replacement pad	S.Y.		
Reinstallation	S.Y.		
Final cleaning	S.F.		
Pad restoration	S.F.		
Mileage out of town	Mi.		
Labor (travel time, furniture disassembly and reassembly)			
Supervisor:	Hr.		
Assistant:	Hr.		
Upholstery			
Base			
Separate cushions	L.F.		
Seat only	L.F.		
Seat and back, ottoman	L.F.		
Full upholstery	L.F.		
Arm covers	Ea.		
Add for Velvet, Haitian cotton, etc.	L.F.		
Add for extra cushions	L.F.		

Item	Unit	Standard Unit Cost	Local Unit Cost
Upholstery (continued)			
Mattresses			
Single	Ea.	_____	_____
Double	Ea.	_____	_____
Queen	Ea.	_____	_____
King	Ea.	_____	_____
Box Springs			
Single	Ea.	_____	_____
Double	Ea.	_____	_____
Queen	Ea.	_____	_____
King	Ea.	_____	_____
Pillows			
Throw	Ea.	_____	_____
Bed	Ea.	_____	_____
Needlepoint	Ea.	_____	_____
Lamp Shades			
Small	Ea.	_____	_____
Medium	Ea.	_____	_____
Large or exotic	Ea.	_____	_____
Special Treatments			
Soil & Stain repellent	L.F.	_____	_____
Deodorization	L.F.	_____	_____
(Or 25% to 30% of above prices)			
Repairs		Quote	_____
Reupholstery (comparable fabric of customer's choice)		Quote	_____
Contractor's overhead & profit, 10% + 10% of charges			_____

Item	Unit	Standard Unit Cost	Local Unit Cost
Draperies			
Base			
1' 0" – 2' 0" (valances)	Pleat	_____	_____
2' 1" – 5' 0"	Pleat	_____	_____
5' 1" – 8' 0"	Pleat	_____	_____
Over 8'	Pleat	_____	_____
Tie backs	Ea.	_____	_____
Add for linings, excessive soil, unusual conditions			
Curtains & Sheers	S.F.	_____	_____
Rehanging			
Curtains	S.F.	_____	_____
Sheers	S.F.	_____	_____
Draperies	Pleat	_____	_____
Valances	Pleat	_____	_____
Remount cornice boards and swags	Window	_____	_____
Window Treatments			
Cornice board	L.F.	_____	_____
Swags	L.F.	_____	_____
Woven Woods	L.F.	_____	_____
Deodorization – 25 – 35% of above charges			
Hard Furniture/Appliances			
Light Residue			
Wipe down and polish horizontal surfaces only		Labor charge_____	
Moderate Residue			
Chair, straight	Ea.	_____	_____
Coffee table	Ea.	_____	_____
End table	Ea.	_____	_____
Rocker	Ea.	_____	_____

Item	Unit	Standard Unit Cost	Local Unit Cost
Hard Furniture/Appliances (continued)			
Bedstead	Ea.	_____	_____
Dresser	Ea.	_____	_____
Chest of drawers	Ea.	_____	_____
Add for mirror	Ea.	_____	_____
Dining room table	Ea.	_____	_____
Appliances	Ea.	_____	_____
Stovetop and oven	Ea.	_____	_____
Washer or dryer, etc.	Ea.	_____	_____
Refrigerator	Ea.	_____	_____
Electronics	Ea.	_____	_____
TV	Ea.	_____	_____
Stereo components	Ea.	_____	_____
Operational Safety Check			
Major applicances & electrical equipment	Ea.	_____	_____
Heavy Residue		Double above	_____
Refinishing		Quote	_____
Dwelling (Cleaning)			
Walls and Ceilings			
Flat latex or enamel	S.F.	_____	_____
Textured	S.F.	_____	_____
Glazed Tile	S.F.	_____	_____
Paneling	S.F.	_____	_____
Clean only	S.F.	_____	_____
Clean and polish	S.F.	_____	_____
Beams, decorative treatments	L.F.	_____	_____

Item	Unit	Standard Unit Cost	Local Unit Cost
Dwelling (Cleaning) – (continued)			
Fixtures (Moderate soot)			
Doors and facings (std.)	Ea.	_____	_____
Windows and frames (std.)	Ea.	_____	_____
Inside only	Ea.	_____	_____
Inside and outside	Ea.	_____	_____
Plate glass	S.F.	_____	_____
Cabinets and vanities (per level)			
Exterior only	L.F.	_____	_____
Exterior and interior	L.F.	_____	_____
Light fixtures (disassemble, clean, and reassemble)			
Standard	Ea.	_____	_____
Chandelier	Ea.	_____	_____
Bath fixtures			
Toilet, sink	Ea.	_____	_____
Tub, shower	Ea.	_____	_____
Vent covers (remove, clean, reinstall)	Ea.	_____	_____
Electrical outlets and switches (remove, clean, and reinstall cover plate)	Ea.	_____	_____
Drapery fixtures	Window	_____	_____
Hard Floors			
Scrub and mop	S.F.	_____	_____
Scrub, mop and wax	S.F.	_____	_____
Strip, seal, refinish	S.F.	_____	_____
Ductwork (seal and deodorize)	Outlet	_____	_____
Miscellaneous Contents			
Pictures and frames			
Framed with glass	Ea.	_____	_____
Prints, photos	Ea.	_____	_____
Oil, acrylics	Ea.	_____	_____

Item	Unit	Standard Unit Cost	Local Unit Cost
Dwelling (Cleaning) – (continued)			
Venetian Blinds			
Single window	Ea.	_____	_____
Add for multiple window (per opening)	Ea.	_____	_____
Bric-a-brac	Item	_____	_____
Packing and Moving			
Labor			
Supervisor:	Hr.	_____	_____
Assistant:	Hr.	_____	_____
Boxes, tape, packing material			
Large	Ea.	_____	_____
Small	Ea.	_____	_____
Deodorization			
Dwelling	C.F.	_____	_____
Contents	25%–35% of charges		_____
Mark-up for Subcontract Work			
Overhead		10%	_____
Profit		10%	_____
Automobile			
Carpet (front or back)		_____	_____
Upholstery		_____	_____
Headliner (dry clean only)		_____	_____

```
                    Quick Repair Contractors
                    1000 Carpenter Drive
                    Atlanta, Georgia 30000
                    (404) 555-1212


                                        April 10, 1990



    Mr. Sam Snead, Adjuster
    XYZ Insurance Company
    2370 Knight Street
    Norcross, Georgia 30071


    Reference:  Contracting Services

    Dear Mr. Snead:

    I am a general contractor specializing in insurance repair work. I
    have been in business for three years in this area, and I believe I
    can provide competent services at competitive prices.

    In fact, I'd like to stop by and discuss a Unit Cost Guide that I
    have prepared to save you time and money in writing estimates.

    I will be in touch with you soon. In the meantime, you may reach me
    at (404) 555-1212. I look forward to meeting you.

                                        Very truly yours,

                                        Mike Jones
                                        President

    MJ:mdc
```

<u>Grantz Heating & Air Conditioning</u>
P. O. Box 100
Sylvan Springs, Florida   10000
(813) 555-1212

August 10, 1990

Ms. Mary Jones
Bucks Insurance Agency
125 Plum Street
Centerville, Florida 10000

Reference: <u>Lightning Damage Claims</u>

Dear Ms.Jones:

   I am a specialist at detecting lightning damage to air
conditioners and other appliances in a building. I know that this
is the subject of many claims against insurance companies, and I
believe many of these claims are paid mistakenly.

   For a flat fee of $85.00, I can conduct a thorough check of an
appliance system and determine the probability of lightning
damage. This method has succeeded in helping insurance companies
save thousands of dollars in unwarranted claims.

   I would very much like the opportunity to discuss my services
with you. I will call you soon to see if your schedule will allow
a brief meeting. In the meantime, please feel free to call me at
(813) 555-1212.

Sincerely,

Jerry Grantz
Proprietor

JG:mdc

<u>Quick Repair Contractors</u>

1000 Carpenter Drive
Atlanta, Georgia   30000
(404) 555-1212

December 2, 1990

Mr. Wilbur Short
123 Hard Street
Duluth, Georgia   30000

Reference: <u>Balance Due on Insurance Repair</u>

Dear Mr. Short:

As general contractors, we performed repairs at your home on
November 10, 1990.

Your insurance company paid its portion of the repair bill, but due
to certain policy conditions, the full amount of repairs was not
paid. Therefore, we are looking to you to pay the balance due of
$250.00.

Please contact me as soon as possible to discuss payment of this
balance.

                                        Very truly yours,

                                        Mike Jones
                                        President

MJ:mdc

<u>Quick Repair Contractors</u>

1000 Carpenter Drive
Atlanta, Georgia 30000
(404) 555-1212

December 22, 1990

Mr. Wilbur Short
123 Hard Street
Duluth, Georgia 30000

Reference: <u>SECOND REQUEST</u>
             <u>Balance Due on Insurance Repair</u>

Dear Mr. Short:

As general contractors, we performed repairs at your home on
November 10, 1990.

This is our second request for payment of the balance owed. As you
are aware, due to policy conditions, your insurance company did not
pay the full amount of repairs. Therefore, we are looking to you
for payment of the balance due, $250.00.

Again, this is our second request. If we do not hear from you in 10
days, we will be forced to refer the matter to a collections
attorney for review.

                            Very truly yours,

                            Mike Jones
                            President

MJ:mdc

Perimeter	Parallelogram	$P = 2l + 2w$
	Rectangle	$P = 2l + 2w$
	Square	$P = 2l + 2w$

Area	Circle	$A = \pi r^2$
	Parallelogram	$A = lw$
	Rectangle	$A = lw$
	Square	$A = lw$
	Trapezoid	$A = 1/2\ h \times (a + b)$
	Triangle	$A = 1/2\ bh$

Circle	Circumference	
	(when diameter is known)	$C = \pi d$
	(when radius is known)	$C = 2\ \pi r$
	Diameter of a circle	
	(when radius is known)	$d = 2r$
	(when circumference is known)	$d = C/\pi$
	Radius of a circle	$r = 1/2\ d$
	Area of a circle	$A = \pi r^2$

Volume	Cylinder	$V = \pi r^2 h$ or $V = bh$
	Solid with a square base	$V = lwh$
	Solid with a rectangular base	$V = lwh$
	Solid with a parallelogram for a base	$V = lwh$
	Triangle or Trapezoid	Surface area of a triangle or trapezoid times the length of the solid.

Pythagorean Theorem	Hypotenuse	$C = \sqrt{a^2 + b^2}$
	Altitude	$C = \sqrt{c^2 - b^2}$
	Base	$B = \sqrt{c^2 - a^2}$

Miscellaneous Formulas	Tank Wall Area	$= \pi\,dh$
	Rafter Length of Gable Roof	$= \sqrt{Rise^2 + Run^2}$
	Sidewall Area of a Room	$= P \times h$

Abandonment
Quoted from insurance policies: *"There can be no abandonment to this company of any property. We need not accept any property abandoned by any insured."* These provisions mean that the insured cannot force the insurer to accept possession of property damaged under a covered loss. That option belongs solely to the insurance company.

Act of God
An event that is the result of an action of natural forces, over which the policyholder has little or no control. Examples are windstorms, floods, earthquakes, and lightning.

Actual Cash Value (A.C.V.)
The cost to replace or restore any item or property at the time of the loss, minus depreciation.

Additional Living Expense
(Coverage D in a Homeowners Policy)
The extra cost of living, above the normal cost, that results from a loss. For example, if a home is damaged by fire so badly that the family cannot live there for three weeks, they would incur extra living expenses from living in a motel, having their laundry done by an outside establishment, eating in restaurants, etc. Additional Living Expense coverage pays for the *EXTRA* expense incurred above what the family normally spends.

Adjuster
A representative of the insurance company who negotiates with everyone involved in a loss, in order to settle the claim equitably. An adjuster deals with the policyholder, repair contractor(s), witnesses, and police officers (if necessary). The adjuster acts as a "middle man" between these parties and the insurance company itself.

Independent Adjuster
An adjuster who is not an employee of any insurance provider, but rather works as a subcontractor to the company or companies and is paid on an hourly basis.

Staff Adjuster
An adjuster who is employed by one particular insurance company.

Public Adjuster
An adjuster who is hired by the policyholder to negotiate with the insurance company, in cases where the policyholder feels his interests are not being served by the staff adjuster or independent adjuster.

Adjustment
The determination of: (1) the cause of a loss and whether it is covered by the policy, (2) the dollar value of the loss, and (3) the amount of money to which the claimant is entitled, after all allowances and deductions have been made.

Apportionment
A formula used to determine how much money will be paid by each

of the policies, when two or more policies cover the same loss.

Appraisal
A dollar estimate of the value of a certain item of property, or the value of a loss. For example, before an underwriter agrees to insure a warehouse, he will obtain an appraisal to see how much it is worth. An appraisal is also needed when a loss occurs. When a claim is made, a contractor may be called in to assess the damages. His assessment is referred to as an *appraisal*.

Appraisal Clause
A clause included in some property policies stating that if there is a disagreement between the policyholder and the insurance company as to the dollar value of the loss, then either party has the right to ask for an appraisal. Both parties hire their own independent appraiser. Then if these two appraisers do not agree, a third appraiser, acting as "umpire" is hired (the expense is shared by the policyholder and the insurance company), and his word is final.

Arson
The deliberate destruction of property by fire. It is often done with the intention of fraudulently collecting the insurance proceeds.

Builder's Risk
A specialized form of property insurance which provides coverage for loss or damage to the work during the course of construction.

Care, Custody and Control
The term used to describe a standard exclusion in liability insurance policies. Under this exclusion, the liability insurance does not cover damage to property that is the responsibility or in the control of the insured.

Certificate
A document issued by an authorized representative of an insurance company, stating the types, amounts, and effective dates of insurance in force for a designated insured.

Claim
A request to be paid for the cost of damages when an insured loss occurs.

Claims Examiner (Supervisor)
The supervisor who oversees the paperwork submitted by the Field Adjuster.

Coinsurance Penalty
The penalty against a policyholder for not carrying enough insurance on his property. In these cases, the full cost of the claim will *not* be paid by the insurance company. Repair contractors should be aware that the policyholder will have to pay a substantial part of the repair bill.

Completed Operations
Liability insurance coverage for injuries to persons or damage to property occurring after an operation is completed (1) when all

operations under the contract have been completed or abandoned; or (2) when all operations at one project site are completed; or (3) when the portion of the work out of which the injury or damage arises has been put to its intended use by the person or organization for whom that portion of the work was done. Completed operations insurance does not apply to damage to the completed work itself.

Comprehensive General Liability

A broad form of liability insurance covering claims for bodily injury and property damage. This insurance combines, under one policy, coverage for all liability exposures (except those specially excluded) on a blanket basis and automatically covers new and unknown hazards that may develop. Comprehensive General Liability Insurance automatically includes contractual liability coverage for certain types of contracts. Products Liability, Completed Operations Liability, and broader Contractual Liability coverages are available on an optional basis. This policy may also be written to include Automobile Liability.

Concurrent Insurance

Insurance under two or more policies which are exactly alike in their terms and conditions, even though they might be different in the dollar amounts of coverage or the dates the policies begin. (Non-concurrent insurance differs from concurrent insurance in the terms and conditions as well.) This condition can complicate the collection process for contractors.

Coverage

An insurance policy, or the dollar protection that policy provides. This amount of protection depends on many factors, including the type of insurance, amount of insurance purchased, policy limits, and exclusions.

Damages

The actual amount of loss sustained by the policyholder. All of a policyholder's damages may not be covered by the policy.

Debris Removal

A clause (usually) included in property policies as an *Additional Coverage*. This clause covers the cost, up to a certain limit, of removing debris that results from a loss.

Deductible

The dollar amount that the policyholder agrees to pay in the event of a loss. The insurance company pays the excess above the deductible, up to the limits of the policy. A homeowners policy usually has a deductible of $250. Commercial buildings may have much higher deductibles; $5,000 to $10,000 deductibles are common.

Depreciation

The dollar amount that is deducted from the value of the item due to age, wear and tear, or obsolescence. Depreciation takes into account the number of years the item probably would have lasted and been useful if the loss had not occurred. Depreciation is deducted from the replacement cost *at the time of the loss, not from the original cost* of the item.

Draft

A note, similar to a check, but differing in that banks will not cash it without certain preliminary processing. A draft takes 10 to 14 days to "clear," or be guaranteed by, the bank. (During this period, the bank sends the draft back to the insurance company for validation.)

Extended Coverage

An endorsement to a property insurance policy which extends the perils covered to include windstorm, hail, riot, civil commotion, explosion (except steam boiler), aircraft, vehicles and smoke.

Independent Adjuster

See **Adjuster**

Independent Agent

A person who sells and services insurance policies of more than one insurance company. The agent must be licensed to represent each of these insurance companies, and earns his income from the commissions of the policies he sells.

Liability

Insurance which protects the insured against liability on account of injury to another person or another person's property.

Lightning

In the context of insurance, a natural event that causes damage. The lightning clause refers only to natural lightning and does not include short circuits or any other damage caused by man-made power sources.

Loss of Use

Insurance protecting against financial loss during the time required to repair or replace property damaged or destroyed by an insured peril.

Loss Payable Clause

A clause in insurance policies protecting the financial institution that holds the mortgage on the insured property. Any payment that the insurance company makes will be made payable to both the policyholder and the lender.

Manufacturers' and Contractors' Liability (M & C)

A form of insurance that covers liability for bodily injury or property damage (except to the contractor's own employees). M & C usually covers claims whether they occur on or away from the contractor's business premises. Coverage is not provided for damage to property that is owned by; rented to; or in the care, custody, or control of the contractor.

Mortgage

A legal pledge of property to a creditor as security for the payment of a loan. A *mortgagee* is the creditor (bank), and the *mortgagor* is the property owner (usually the policyholder).

Mortgage(e) Clause

See **Loss Payable Clause.**

Personal Property

In the context of insurance, movable property, as opposed to real property (land and buildings).

Personal Injury

Bodily injury, and also injury or damage to the character or reputation of a person. Personal injury insurance includes coverage for injuries or damage to others caused by specified actions of the insured, such as false arrest, malicious prosecution, willful detention, imprisonment, libel, slander, defamation of character, wrongful eviction, invasion of privacy, or wrongful entry.

Personal Liability

See **Personal Injury**

Policy

The written contract between an insurance company and the policyholder. The policy contains the details of the insurance agreement, including any modifying endorsements (additions) attached either at the time of issue or later.

Property Insurance

Insurance that compensates the insured for the loss of property (real or personal) resulting from direct physical damage.

Real Property

Real estate (land and/or buildings) as opposed to movable (personal) property.

Replacement Cost Coverage

A type of insurance that guarantees that the insurance company will pay to replace the damaged property with new property (depreciation will not be deducted). *See* **Actual Cash Value.**

Salvage

The remaining value of property after it has been partially damaged. Salvage can also mean the act of saving and preserving such partially damaged property, so as to prevent further loss. For example, a flooded carpet can be salvaged by drying and cleaning it.

Settlement

The total amount of money that both the insurance company and the policyholder agree on, to close the claim. (The settlement may not cover the full cost of the damage, because of deductibles, coinsurance, etc.)

Smoke Damage

Damage that is a result of a fire, or that is caused by "sudden, unusual, and faulty" operation of a heating or cooking unit that is connected to a chimney by a smoke pipe or vent.

This kind of smoke damage is only covered if it is sudden, unusual, and accidental; gradual smoke leaks, for example, would not be covered.

Special Agent
A salaried employee of an insurance company who acts a "middle man" between the company itself and its many local sales agents. There is just one Special Agent for a given territory, although there may be several sales agents. The Special Agent does not usually have any contact with policyholders themselves.

Special Hazards
Insurance coverage for damage caused by additional perils or risks. Such a clause is included in the property insurance at the request of the contractor or at the option of the owner. Examples are sprinkler leakage and water damage.

Underwriter
An insurance company or an employee of the insurance company who is responsible for approving the issue of policies. Underwriters decide on the amounts, the terms, and the conditions of the policies.

Valued Policy Laws
State laws governing the amount of insurance companies must pay relative to the value of the loss. Practically all Standard Fire Policies limit claim payments to the "actual cash value" of the property at the time of loss, even though the full amount of insurance might be more. (If a house is insured for $100,000 but it is really only worth $80,000 the insurance company would only have to pay $80,000 if the house burned to the ground.) However, in some states there are "Valued Policy Laws" which require the insurance company to pay the full amount of insurance. (In those states, the homeowner would get the full $100,000.)

INDEX